Fluency Doesn't Just Happen in Multiplication and Division

Fluency in math doesn't just happen; it is a well-planned journey. In this book, you'll find practical strategies and activities for teaching your elementary students basic multiplication and division. The authors lay out the basic framework for building math fluency using a cycle of engagement (concrete, pictorial, abstract) and provide a multitude of examples illustrating the strategies in action.

You'll learn how to help students to model their thinking with a variety of tools; keep students engaged through games, poems, songs, and technology; assess student development to facilitate active and continuous learning; implement distributed practices throughout the year; and boost parental involvement so that students remain encouraged even as material becomes more complex.

A final chapter devoted to action plans will help you put these strategies into practice in your classroom right away. Most importantly, you'll open the door to deep and lasting math fluency.

Dr. Nicki Newton has been an educator for 35 years, working both nationally and internationally with students of all ages. She has worked on developing Math Workshop and Guided Math Institutes around the country; visit her website at www.drnickinewton.com.

Ann Elise Record has been an educator for over 22 years as a teacher, math coach and specialist, adjunct faculty member for Plymouth State University, and, currently, as a national speaker and independent elementary math consultant. She can be found on Facebook (www.facebook.com/groups/2022849141262766) facilitating the Math Running Records group. Visit her website at www.anneliserecord.com.

Dr. Alison J. Mello has been in education for nearly 30 years as a classroom teacher, math specialist, director of curriculum, and assistant superintendent. Alison now enjoys serving teachers through her work as a math consultant, graduate instructor, and national speaker. You can find her at alisonmellomathconsulting.com.

Also Available from Dr. Nicki Newton
(www.routledge.com/eyeoneducation)

Fluency Doesn't Just Happen with Addition and Subtraction:
Strategies and Models for Teaching the Basic Facts
With Alison Mello and Ann Elise Record

Guided Math Lessons in Kindergarten:
Getting Started

Guided Math Lessons in First Grade:
Getting Started

Guided Math Lessons in Second Grade:
Getting Started

Guided Math Lessons in Third Grade:
Getting Started

Guided Math Lessons in Fourth Grade:
Getting Started

Guided Math Lessons in Fifth Grade:
Getting Started

Day-by-Day Math Thinking Routines in Kindergarten:
40 Weeks of Quick Prompts and Activities

Day-by-Day Math Thinking Routines in First Grade:
40 Weeks of Quick Prompts and Activities

Day-by-Day Math Thinking Routines in Second Grade:
40 Weeks of Quick Prompts and Activities

Day-by-Day Math Thinking Routines in Third Grade:
40 Weeks of Quick Prompts and Activities

Day-by-Day Math Thinking Routines in Fourth Grade:
40 Weeks of Quick Prompts and Activities

Day-by-Day Math Thinking Routines in Fifth Grade:
40 Weeks of Quick Prompts and Activities

Leveling Math Workstations in Grades K–2:
Strategies for Differentiated Practice

Daily Math Thinking Routines in Action:
Distributed Practices Across the Year

Mathematizing Your School:
Creating a Culture for Math Success
Co-authored by Janet Nuzzie

Math Problem Solving in Action:
Getting Students to Love Word Problems, Grades K–2

Math Problem Solving in Action:
Getting Students to Love Word Problems, Grades 3–5

Math Workshop in Action:
Strategies for Grades K–5

Math Running Records in Action:
A Framework for Assessing Basic Fact Fluency in Grades K–5

Math Workstations in Action:
Powerful Possibilities for Engaged Learning in Grades 3–5

Fluency Doesn't Just Happen in Multiplication and Division

Strategies and Models for Teaching the Basic Facts

Dr. Nicki Newton, Ann Elise Record, and Dr. Alison J. Mello

Routledge
Taylor & Francis Group

NEW YORK AND LONDON

Designed cover image: Miguel Alfau

First published 2024
by Routledge
605 Third Avenue, New York, NY 10158

and by Routledge
4 Park Square, Milton Park, Abingdon, Oxon, OX14 4RN

Routledge is an imprint of the Taylor & Francis Group, an informa business

Library of Congress Cataloging-in-Publication Data
Names: Newton, Nicki, author. | Record, Ann Elise, author. | Mello, Alison J., author.
Title: Fluency doesn't just happen in multiplication and division : strategies and models for teaching the basic facts / Dr. Nicki Newton, Ann Alise Record, and Dr. Allison J. Mello.
Description: New York, NY : Routledge, 2024. | Includes bibliographical references.
Identifiers: LCCN 2023054619 (print) | LCCN 2023054620 (ebook) | ISBN 9781032614236 (hbk) | ISBN 9781032557007 (pbk) | ISBN 9781032614229 (ebk)
Subjects: LCSH: Arithmetic—Study and teaching (Elementary) | Multiplication—Study and teaching (Elementary) | Division—Study and teaching (Elementary)
Classification: LCC QA135.6 .N4854 2024 (print) | LCC QA135.6 (ebook) | DDC 372.7/2—dc23/eng/20240209
LC record available at https://lccn.loc.gov/2023054619
LC ebook record available at https://lccn.loc.gov/2023054620

ISBN: 978-1-032-61423-6 (hbk)
ISBN: 978-1-032-55700-7 (pbk)
ISBN: 978-1-032-61422-9 (ebk)

DOI: 10.4324/9781032614229

Typeset in Berling and Futura
by Apex CoVantage, LLC

Nicki: I dedicate this book to Mom and Pops, always.

Ann Elise: I dedicate this book to my sister, Janice, my brother, Justin, and my parents-in-law, Burgess and Sandy, for their endless love and support.

Alison: I dedicate this book to my Foxborough family and all of the teachers and leaders there who continue to support and inspire me. I love you all.

Contents

Foreword

Math Is Not a Spectator Sport . . .

Kids need to play the game, not watch it!

As I read through this book, that line hit me. All four of my kids play sports, and I couldn't imagine their time learning a sport being where they sit in a seat and listen to their coach talk about how to do it.

Or watch their coach play and just take notes.

They all learned by doing.

Yes, the coach was there giving guidance along the way, but it was while they were doing. They were immersed in playing (and practicing).

We let children dive right into playing a sport, even though they may have never done it. They aren't perfect. They do things incorrectly, and the coach is there to give instruction in the moment.

You can tell kids how to run plays in a game, but it doesn't sink in until they are in the moment actually doing it.

We need to let kids have that same experience when developing their math understanding. Kids need to dive in and play. As they explore by doing, we can instruct and give guidance.

If you are like me, I learned my math facts through memorizing them. I read them over and over like a sentence until it was ingrained in my memory:

$3 \times 4 = 12$
Three times four equals twelve.
Three times four equals twelve.
Three times four equals twelve . . .

There was no connection to other facts, well except for 4×3. Other than that, they were isolated facts. It worked for me throughout school because I was a good memorizer. To this day, I know all my math facts.

It worked. Until it didn't.

It wasn't until I was an adult that I realized I didn't really understand math. I could do math and I knew my "facts," but I didn't understand it.

I was mindlessly following rules and procedures. Even the regurgitation of facts was mindless. It was just a piece of information for me to spout out. It didn't mean anything.

Three times four equals twelve.

It only meant that.

It did NOT mean 3 groups of 4. Or 4 groups of 3. I didn't work with drawing pictures or using manipulatives to model. I didn't have a visual of what it meant. And I for sure wasn't thinking about how it connects to other things I might know (like 2 groups of 3 or 2 groups of 4). All I knew was *three times four equals twelve*.

Understanding math is like swimming, while memorizing math is like floating. To swim is to be active, digging through the water to get to your destination. Floating is passive; you can get to the destination, but there is no real struggle involved, no muscle memory that gets formed.

Knowing math facts is still very important. But how we help kids get to that destination needs a serious upgrade. This book is just that.

Dr. Nicki, Ann Elise, and Alison walk you through how to help your students get to the point that they know their multiplication and division facts from memory, not through repeated drills and memorization but in a way that ensures your students are active participants and not passive spectators.

Your students will learn how connected mathematics is as they jump in and play with manipulatives and visuals that help them become fluent and flexible thinkers.

The way to solidify information into our long-term memory, and quickly retrieve it when needed, is to be active participants in the collection of the information AND build as many connection points to that information as possible.

Enjoy this book, and I know your students will enjoy playing and building connections as they work their way to building fluency with multiplication and division facts.

Christina Tondevold

Preface

I am so excited to be writing the second volume of this great book with two extraordinary colleagues. We all share a deep love for the teaching of math. We are especially interested in the teaching and learning of basic fact fluency. In this book, we will focus on developing a thorough understanding of multiplication and division basic facts. We applaud those who have trailed a path before us, and now we would like to further the discussion by unpacking more activities that build conceptual understanding, work on procedural fluency, and ask students to reason, develop strategic competence, and continue to work toward a productive disposition (National Research Council, Mathematics Learning Study Committee, 2001). We also want to consider the ways to meaningfully integrate technology. We have divided this book into 9 chapters:

Chapter 1: Introduction

In Chapter 1, we discuss the research about teaching students basic facts. We specifically focus on the frameworks for teaching multiplication and division facts. Dr. Nicki Newton calls them the "Dolch words of math" (Newton, 2010, personal communication). Dolch words are all the basic words that students need to know to become fluent readers. In the same way, the math Dolch words are the facts that students need to know so that they can do the higher-level mathematics without being bogged down in the basic fact stuff.

Chapter 2: Modeling Math Facts

In this chapter, we are going to delve deep into what it means and how to teach multiplication and division facts using various models. We will discuss:

- The classroom as a toolkit
- Individual toolkits
- Teacher toolkits
- Digital toolkits

We explain the importance of how tools help students to model their thinking. Tools help students to understand the math they are doing.

Chapter 3: Exploring and Learning Multiplication Facts

In this chapter, we are going to delve deep into what it means and how to teach multiplication facts within 100. We will discuss teaching:

- 0 and 1
- 10 and 5
- 2, 4, 8
- 3, 6, 9
- 7

We explain a cycle of engagement that looks at teaching and learning using concrete, pictorial, and abstract representations. We will talk about which activities to use and how to develop deep levels of understanding. We will also discuss picture books, songs, poems, videos, and spotlight activities for teaching these facts.

Chapter 4: Exploring and Learning Division Facts

In this chapter, we are going to delve deep into what it means and how to teach division facts within 100. We will discuss teaching:

- Dividing by 1
- Dividing 0 by a number
- Dividing a number by itself
- Dividing a number by 2 (half)
- Dividing a number by its half
- Thinking multiplication

We discuss a cycle of engagement that looks at teaching and learning using concrete, pictorial, and abstract representations. We will talk about which activities to use and how to develop a deeper level of understanding. We will also discuss picture books, songs, poems, videos, and spotlight activities for teaching these facts.

Chapter 5: Multiplication and Division Problem Types

In this chapter, we will discuss research-based problem structures.

Multiplication and Division Types

- Arrays
- Equal groups
- Multiplicative comparison

We discuss a cycle of engagement that looks at teaching and learning using concrete, pictorial, and abstract representations of the various story types. We will talk about which activities to use and how to develop a deeper level of understanding. We will talk about the different levels of word problems and where to get resources to teach them.

In the next part of the book, we will explore how to assess, engage in ongoing practice throughout the year, involve parents, and make and implement a fluency plan of action.

- **Assessment**
- **Daily fluency routines**
- **Parents**
- **Action planning**

Assessment is the linchpin of any fluency plan. We must know where the students are starting and where they end each year. We have to have an articulated plan of action that includes the parents, the students, and other staff. This can all happen with a well-mapped-out plan for success.

Chapter 6: Number Flexes for Daily Math Fluency Practice

In this chapter, we will discuss engaging activities that can be used to distribute practice across the year rather than just massed in a unit of study.

- What doesn't belong?
- Over, under, the same
- True or false
- Splat/power of color
- Number talks
- Number strings
- I was walking down the street
- What's the model
- Model that
- What's tricky?

Energizers and routines help students to practice in wonderful ways so that they maintain their skills, strengthen them, and develop new ones. Many students even like to make up their own routines.

Chapter 7: Assessment

In this chapter, we look at the many different aspects of assessment that take place along the way as students learn their facts. We will explore:

- Pre-assessment
  - Math Running Records
  - Two-colored probe
- Ongoing assessment
- Posters

- Quizzes
- Entrance and exit slips
- Summative assessments

We want to emphasize that the assessment of learning facts should be a scrapbook rather than a snapshot of what students know how to do and what they need to learn next (Wiggins, 2012).

Chapter 8: Parental Involvement

In Chapter 8, we discuss the crucial element of home support. We talk about how to set parents up to be able to help their students with their basic math facts.

- Letters
- Game boards
- Flashcards
- Online games

We strongly believe that if parents or guardians know how to help their students, then they will.

Chapter 9: Action Plan

In this chapter, we discuss making and doing an action plan. If you plan and write it down, then you are much more likely to get started and continue doing it. So, we help you navigate some of the key elements of making a plan that you can use.

- Practice
- Assessment
- Fluency map

It is with great joy that we have written this book to continue an ongoing discussion about teaching students their basic multiplication and division math facts. We wish you many "Aha!" moments along the way.

REFERENCES

National Research Council, Mathematics Learning Study Committee. (2001). *Adding it up: Helping children learn mathematics*. National Academies Press.

Wiggins, G. (2012). Seven keys to effective feedback. *Feedback, 70*(1), 10–16.

Acknowledgements

Writing a book is a wild adventure. I am so glad we took this adventure together. Ann Elise and Alison have been wonderful writing companions. Anna and Gloria help me with the day-to-day of it all. My friends encourage me on the phone to do the work. My family is always there—my brother, Marvin, my sister, Sharon, my Tia, Mary, my cousin, Clinese, and my nieces, nephews, cousins, and godchildren that support me along the way. Also, my Uncle Bill (now 92) is still encouraging me and just as proud as ever of every book I write. I pay tribute to all the teachers and students who show me the path as I walk it with them. They are all so kind and generous. They teach me well. This book could not have been written without editor extraordinaire Lauren. We have written about 20 books together and she inspires me, encourages me, and kicks me along to great things every time. I'm blessed to have her on this journey. The entire team of copyeditors, graphic designers, and production staff that helped put the book together have done a phenomenal job, as always. I thank them for their patience. Mostly, I thank God for every breath I take, every thought I think, and every word I write.

Nicki

What an honor and privilege to have been on this book-writing journey again with Dr. Nicki Newton and Dr. Alison Mello! I so love learning with and from them! First and foremost, I'm thankful for my incredible husband, Dan, who is my biggest cheerleader and who provides extra cuddles to the pets while I'm traveling the country. Our children, Matthew and fiancé Hannah, Kathryn and husband Bryan, and our bonus grandchildren Hank, Aiden, and Jackie are my heart and I'm always counting down the next trip to Madison, WI, to visit with them. I am eternally grateful for all my friends and family who provide endless love and support as I continue this crazy wonderful journey as a math consultant. The treasured weekly video chats, texts, emails, shared Tik Toks, and Marco Polo videos provide endless love and laughter every day and are particularly welcome while I'm traveling. I am so thankful to have the wonderful opportunity to view life and math through the eyes of my favorite young people—my three (so far . . .) grandchildren, my niece, Kylynn, and friends Delia and Leland. I also send out my thanks for our editor, Lauren, and everyone on the Routledge team that had a part in taking our Google Documents and transforming them into the book in your hands. Last and certainly not least, I will be forever grateful to Dr. Nicki Newton for welcoming me yet again to write this book with her and Dr. Alison Mello. I am awed by the extent of her knowledge of research, her generosity to others, and her continual sense of wonder at the way children learn and grow. She is my math mentor and, I'm so thankful to say, my close friend. Here's to many more adventures!

Ann Elise

I can honestly say that I never aspired to write a book, let alone multiple books. If not for my amazing friend, Dr. Nicki Newton, I doubt that I would be here, but anyone who knows her knows that she is tenacious . . . and she made it happen! She continues to see in me things that I don't always see in myself, and I am forever grateful for her support, inspiration, and stubborn persistence! She starts nearly every conversation with the phrase "We should write a book about . . .," and I have come to realize that if she says we will, there is no changing her mind! I am so grateful that she insisted that I partner with her and Ann Elise to bring this book to life. It was an honor. I am also grateful for the opportunities that I have every day to meet and work with amazing and dedicated teachers. Most of all, I am grateful for my husband, Jon, our incredible daughters, Alaina and Julia, and our families who always support me even when writing takes me away from them. I could never do what I do without their support, and I am deeply aware of how blessed I am to have them. Lastly, I want to give a special shout out to my mother-in-law, Marianne, and my best friend LeeAnn, two strong, positive, amazing women who always want to hear every detail of what I am doing and are among my biggest cheerleaders. I love you both.

Alison

CHAPTER 1

Introduction

Do not be content with the right answer. Always demand explanation.

(Van de Walle, 2001, p. 425)

INTRODUCTION

Fluency doesn't just happen! It is a well-planned journey. This book is part 2 in a series meant to help you navigate that journey. Fluency is a multi-dimensional concept. We like to think of it as a four-legged stool: accuracy, flexibility, efficiency, and strategy selection. Ann Elise once said that "automaticity has hijacked fluency." How many students do you know who can say their facts from memory but have no number sense? So, if you say, "What's 7 × 8?", they might say 56, but if you say, "What's 7 ´ 9?", they say they don't know because they only know the poem for 7 × 8 . . . "5, 6, 7, 8 . . . 7 × 8 is 56." There you have it. An answer with no understanding!

This book is about scaffolding fluency with multiplication and division so that students have a profound sense of numbers. We want them to be able to think about numbers in a variety of ways. The best way to develop fluency with numbers is to develop number sense and to work with numbers in different ways, not to blindly memorize without number sense (Boaler, 2015). Parrish, drawing from Fosnot and Dolk (2001), defines fluency as "knowing how a number can be composed and decomposed and using that information to be flexible and efficient with solving problems" (Parrish, 2010, p. 159). If students have a strong foundation with addition and subtraction, then continuing that same line of thought with multiplication and division should not be that difficult. They should be used to breaking numbers apart and putting them back together as well as thinking about numbers in relation to other numbers.

The focus of this book is to provide a framework for teaching and learning multiplication and division facts through a variety of engaging, ongoing, interactive, rigorous, student-friendly activities that build a fundamental understanding of how numbers are in relationships with each other. The research resoundingly states that computational fluency is multi-dimensional (speed and accuracy, flexibility and efficiency) (Brownell & Chazal, 1935; Brownell, 1956/1987; Kilpatrick et al., 2001; National Council of Teachers of Mathematics, 2000). Understanding should be grounded in place value, properties, and the relationship between the operations (NCTM, 2000).

DOI: 10.4324/9781032614229-1

DOLCH WORDS OF MATH

Students should *learn* their facts rather than *memorize* them. If you just memorize them, then you can easily forget them. If you learn them, then you can always access them through a variety of strategies based in place value, properties, and the relationships between the operations. There is a continuum for learning basic facts. Baroody (2006) calls it the "Phases of Mastery." Battista (2012) calls it the "Levels of Sophistication." This continuum has been discussed by many researchers. Basic facts for multiplication and division are products and quotients within 100. Here is a suggested method of teaching the facts that is based on number relationships (see Figure 1.1).

Multiplication	Division
0 and 1	Dividing by 1
10 and 5	Dividing 0 by a number
2, 4, 8	Dividing a number by itself
3, 6, 9	Dividing by 2
7	Dividing a number by its half
Review Squares	Thinking Multiplication

FIGURE 1.1 Dolch Words of Math

It is vital that students learn their facts because this lays the foundation for more complex math skills such as fractions, decimals, and algebra (Flowers & Rubenstein, 2010). By the end of third grade when students fully transition from additive to multiplicative reasoning, they are wrapping up their learning of "the four basic operations," making this a very critical time. One outcome of this grade level is that students are expected to be fluent with all of their basic facts. As O'Donnell and SanGiovanni note, 4th grade doesn't spend a great deal of time learning the basic facts, yet we know that many students are still very shaky on their basic facts in 4th and even 5th grade. I always tell 5th grade teachers that this is a 911 situation; whatever you do, you have to make time somewhere in your program to develop mastery of the basic facts because middle school teachers will not do it. This means that if students leave 5th grade without the knowledge and skills of their basic facts, they will probably never learn them and be handicapped throughout middle school, high school, and even college.

The research tells us that students should practice in a variety of engaging ways over time, so that they acquire their facts with meaning (Stickney et al., 2012; Boaler, 2015; Van de Walle, 2007). Data from quality assessments (sorry . . . timed tests don't qualify!) should be on hand before students head off to practice. It is essential to assess all of your students in a meaningful way so that you can find out where they are on the learning trajectory of the facts. Having this information will empower you to offer opportunities for targeted, differentiated practice within the students' zone of proximal development so that they reach their grade-level designated fluency.

STRATEGY TALK

As students are learning their facts, there are different approaches to working with numbers. These strategies are often indicative of where students are on the developmental continuum. Each one has a specific name (see Figure 1.2).

Strategies				
Counted All	**Skip Counting**	**Derived Fact**	**Hybrid Strategies**	**Automatic Facts**
When students are counting all they start at one and count every number. 3 x 4 Sounds like: 1, 2, 3, 4, 5, 6, 7, 8, 9, 10, 11, 12	Students skip count. 3 x 4 sounds like: 4, 8, 12	Students use facts they know... so for example 6 x 7 3 x 7 = 21 doubled makes 42	Students mix strategies. 6 x 7 6 x 5 = 30 36, 42	This is when students know their facts without having to think about them. Logan calls this the "instant popping into of mind." (1991)

FIGURE 1.2   Cycle of Engagement: Concrete, Representational, Abstract

It is important to use manipulatives when students are learning multiplication and division. Oftentimes, students are rushed to practice these facts at an abstract level with flashcards and board games without being given the time to conceptually understand what they are doing and how these operations are related. Many researchers maintain that learning through the cycle of engagement of concrete, representational, and abstract gives students access to deeper understanding of mathematical concepts (Anstrom, 2006; Bender, 2009; Devlin, 2000; Van de Walle, 2001; Maccini & Gagnon, 2000). This cycle is a three-step instructional process that allows students to gain a conceptual understanding of a strategy by working with manipulatives. The second part of the cycle moves students from the objects to pictorial representations of the equation. The third part of the cycle moves students to an abstract level with the concepts by representing the operation with numbers and symbols. Each part of the cycle should be connected with the other parts of the cycle. It is not a hierarchy, nor is it linear. Students should understand the relationships between the parts. This doesn't mean that you necessarily always do all three representations at the same time. Sometimes you may be focused on one part of the cycle. This understanding will develop over time as the students connect these various representations.

For example, if students are learning about equal groups for multiplication, they could act out the word problems: Sue had 3 bags with 5 marbles in each bag. First, the student would get 3 bags and put 5 marbles in each one. Next, the student would draw a picture to represent this. Finally, the student would write the multiplication sentence. This would happen all in the same lesson with students talking about the connections between the different representations. These are concept-building lessons. Although you don't do all of these forever, we would argue that if you start at this level and build a strong conceptual understanding, students will more deeply understand multiplication and division (see Figure 1.3).

Concrete Activities	Pictorial Activities	Abstract Activities
One-inch tiles	Pictures	Board Games
Cubes/Bears	Tally marks	Dice Games
Playdoh	Stickers	Domino Games
Beans	Number Frames	Flashcards/Card Games

FIGURE 1.3 Concrete, Pictorial, and Abstract Activities

Another example is practicing at these levels to build understanding of the properties. Teachers often tell students that 9×1 is the same as $1 \, ' \, 9$. While it may result in the same product, it is actually a very different situation. Let's look at this in a contextualized way to better understand the nuances of the difference. Imagine you own a boat company, and a person asks you for 1 boat that fits 9 people. If they arrive and you present them with 9 boats that fit 1 person each, they will likely be quite upset because that is not what they asked for! Keep this in mind as you have students solve problems to highlight that reasoning about the math they are doing is essential.

As Van de Walle, 2007 noted, "Basic facts are the linchpin of learning mathematics because students will build on this foundational knowledge to do everything else in math. Furthermore, 'All children are able to master the basic facts—including children with learning disabilities'" (2007, p. 24).

MAKING THE CONNECTION BETWEEN MULTIPLICATION AND DIVISION

21ST CENTURY COMPONENTS

The students that we teach are called "digital natives" (Prensky, 2001). We are called the "digital immigrants." They were born into a world where the internet is ever present. They have learned much of what they know from the internet. How can we leverage this? We know that we must incorporate the teaching and learning of the basic facts into our daily routines. This means that one way they should practice their basic facts is using internet games (although it is important to note that most of these games are working mainly on accuracy and instant recall). Another way to use technology is as a tool for students to show what they know. They can make podcasts, screen casts, videos, and narrated Power-Points to show what they know. In this way, they shift from consumers of technology to producers, which is at the apex of Bloom's taxonomy (Anderson et al., 2001).

PURPOSEFUL PRACTICE

The brain researchers concluded that automaticity should be reached through understanding of numerical relations, achieved through thinking about number strategies.
(Delazer et al., 2005 cited in Boaler et al., 2015)

Basic math facts should be reinforced in a variety of ways. Stille found that when "basic multiplication facts were introduced through the use of strategies that included exploration, manipulative objects, guided reinvention, developmental pacing, interactive dialogue and relational understanding with the goal of building fluency in the facts," this helped students to understand what they were doing and to reason about numbers better (2017). Practice should be meaningful. Students should practice number combinations in ways that build understanding and familiarity. Over time, students become automatic with various number combinations. Van de Walle (2007) warns us that we shouldn't just be drilling students but rather only after they have learned efficient strategies should they work on drills. I think this is more about having students work toward automaticity. After students understand the concept and are comfortable with the number combinations, they then can play games that work on instant recall.

Students do not have to do drills to work on instant recall; playing board games, card games, dice games, and domino games helps students work on instant recall as well. Differentiating workstations addresses the idea that meaningful practice should be individualized (Van de Walle, 2007).

Importantly the study also found that those who learned through strategies achieved "superior performance" over those who memorized, they solved problems at the same speed, and showed better transfer to new problems. Boaler et al. (2015)

When students are working in their zone of proximal development (Vygotsky & Cole, 1978), then they are not bored or frustrated. They can actually build a solid foundation of understanding. This specifically looks like some students working on multiply by 4 facts and other students working on multiply by 9 facts. Everyone is working toward the grade-level standard but at their own pace. In literacy, we would never say that everyone has to read the same books at the same time, but in math, we do exactly that. It doesn't work for literacy, and it doesn't work for math either. Students will progress toward the grade-level standard at different times, and this is normal and expected, so therefore we must plan for it!

Part of purposeful practice entails students self-monitoring. There are different ways to do this. Students might have a fact fluency folder or fact practice rings (see Figure 1.4).

Fact Fluency Folder	Fact Practice Rings

FIGURE 1.4   Ways to Practice

ENGAGING APPROACHES

Our students are diverse learners. Building on learning preferences and ideas about students engaging in learning through a variety of intelligences, we suggest an array of ways to engage students in learning their basic facts (Stille, 2017; O'Connell & SanGiovanni, 2011). In addition to digital and traditional games, students can also practice through poems, songs, picture books, and videos. Not all of these options offer the same impact, but they do appeal to different types of learners. Be cautious not to employ any activities that rely on mnemonic devices that teach students isolated facts but don't build number sense. We don't recommend that you use those! However, there is some really good, catchy, fun stuff out there that builds number sense. We highly recommend that you use that!

Throughout the chapters, we will look at how literature can serve as a springboard into unpacking various strategies. We will look at different ways that students can work on their facts, from individual games to partner games and small-group games. We will also discuss how various energizers and routines can help students to engage in distributed practice over time.

MATH LITERATURE

Children's picture books and poems can be a great springboard into teaching and learning the basic math facts. They provide an understandable context so students can follow the story. They often intentionally teach a skill or concept. They are engaging and have many extension possibilities. There is a great deal of research supporting the idea that using literature to teach math can be very powerful (Whitin & Wilde, 1992, 1995; Whitin & Whitin, 2004; Burns, 1992, 1995; Zambo, 2005). Scholars note that stories help students to understand the math and construct meaning through the story (Whitin & Whitin, 2004; Burns, 1995). Draper (2002) noted that "literacy and literacy instruction are necessary parts of mathematics instruction" (p. 523). Math literature "allows students to think deeply, discuss, debate" (Ducolon, 2000) and "gain experience with solving word problems in familiar stories" (Ward, 2005).

> Integrating literature within mathematics lessons not only develops literacy skills, but also promotes mathematical language and problem solving. Moreover, the visual representation in the literature books not only stimulates readers, but also provides informative story lines that foster children's curiosity. The power of children's literature and mathematics provide readers with opportunities to try different strategies and to scaffold previous experiences to broaden one's learning.
>
> Wilburne and Napoli (2008)

Many children's books present interesting problems and illustrate how other children solve them. Through these books, students see mathematics in a different context while they use reading as a form of communication. NCTM (1989) (p. 28)

GAMIFICATION

Games are a great vehicle to get students to practice their facts. Research shows that students will spend 100 hours getting good at a video game (Gee, 2005). There are certain elements of game theory that students connect with. After writing extensively about game theory and education, Gee said:

> So the suggestion I leave you with is not "use games in school"—though that's a good idea—but: How can we make learning in and out of school, with or without using games, more game-like in the sense of using the sorts of learning principles young people see in good games every day when and if they are playing these games reflectively and strategically?
>
> (p. 11)

Dr. Nicki calls it the *purple popsicle stick theory*. She notes that, "If you give students something that they need to do but don't want to do on a worksheet, there is resistance. But, if you put it on a purple popsicle stick and you tell students that whoever gets 5 first wins, they are all in!" The point is that students *love* games. They *want* to play games. Teachers just have to offer academically rigorous, standards-based games that are engaging, and these do not need to involve lots of prep or heavy lifting. Find something that works and stretch it out. Students love bonus rounds! When students love what they are doing they will stay engaged in what is called "infinite play," which means that they will play a game over and over until they master it because they are intrigued by it (Knewton, n.d.).

Why do games work? Students know where they are, what they are trying to do, what the end game is, and what they need to do to get there. They engage and reflect. They monitor their progress. They do all the stuff we want them to do in school! Gee (2005) notes that students take on a particular identity in a game.

- Commitment
- Interaction
- Production
- Risk taking
- Customization
- Agency
- Well-ordered problems
- Challenge and consolidation
- "Just in time" or "on demand"
- Pleasantly frustrating—idea of doable but challenging
- Performance before competence

Hello! Are we listening? How do we as educators "up our game?" Imagine if students worked on their facts with the engagement of a video game! Imagine a classroom where, when you tried to stop math class, the students begged to play for just a few more minutes! Imagine a space where everyone knew exactly where they were on the learning continuum and what they needed to do to get better and they had a plan! This can happen. Hopefully, with some of the suggestions in this book, we can get closer to creating those spaces.

KEY POINTS

- Dolch words of math
- Cycle of engagement
- 21st century technologies
- Engaging approaches
- Math literature
- Gamification

SUMMARY

Facts are the foundation for most of the math that students will learn in school. There is a specific developmental approach to teaching facts. We all need to learn the sequence and assess according to it so that we can address the specific needs of our students. The cycle of engagement builds a strong understanding of the facts by having students work concretely, pictorially, and abstractly on concepts and connecting those representations when possible. It is important to integrate 21st-century technologies into the teaching and learning of facts so that students can practice in engaging and rigorous ways. We must be ever cognizant of applying multiple approaches so that we tap into the different ways that students learn. We need to consider the implications of game theory on our own pedagogy to see how we might better capture and keep the attention of our students.

CALL TO ACTION

 1. Share your favorite "Aha!" moment from this chapter on social media to help spread the movement! #FDJH

 2. Take a photo of your student's fluency folders and share on social media! #FDJH

3. Here are some fact fluency cover ideas so you can get started today!

REFERENCES

Anderson, L. W., Krathwohl, D. R., & Bloom, B. S. (2001). *A taxonomy for learning, teaching, and assessing: A revision of Bloom's taxonomy of educational objectives* (Complete Edition. New York: Longman)

Anstrom, T. (2006). Supporting students in mathematics through the use of manipulatives. *American Institutes for Research, 1,* 15.

Baroody, A. (2006). Why children have difficulties mastering the basic number combinations and how to help them. *Teaching Children Mathematics, 13*(1), 22–31.

Battista, M. (2012). *Cognition-based assessment and teaching of addition and subtraction: Building on students' reasoning.* Heinemann.

Bender, W. (2009). *Differentiating math instruction: Strategies that work for K-* classroom.* Corwin Press.

Boaler, J. (2015). *Fluency without fear.* Retrieved May 15, from https://www.youcubed.org/evidence/fluency-without-fear/

Boaler, J., Williams, C., & Confer, A. (2015). *Fluency without fear: Research evidence on the best ways to learn math facts.* youcubed@Standford University.

Brownell, W. A. (1987). AT classic: Meaning and skill—maintaining the balance. *Arithmetic Teacher, 34*(8), 18–25 (Original work published 1956).

Brownell, W. A., & Chazal, C. B. (1935). The effects of premature drill in third-grade arithmetic. *The Journal of Educational Research, 29*(1), 17–28.

Burns, M. (1992). *Math and literature: (K-3). Book one.* Math Solutions Publications.

Burns, M. (1995). *Writing in math class.* Math Solutions Publications.

Devlin, K. (2000). Finding your inner mathematician. *The Chronicle of Higher Education, 46,* B5.

Draper, R. (2002). School mathematics reform, constructivism and literacy: A case for literacy instruction in the reform-oriented math classroom. *Journal of Adolescent & Adult Literacy, 47,* 520–529.

Ducolon, C. (2000). Quality literature as a springboard to problem solving. *Teaching Children Mathematics, 6*(7), 442–447.

Flowers, J., & Rubenstein, R. (2010). Multiplication fact fluency using doubles. *Mathematics Teaching in the Middle School, 16*(5), 296–301.

Fosnot, C. T., & Dolk, M. (2001). *Young mathematicians at work: Constructing multiplication and division.* Heinemann.

Gamification. Retrieved November 20, 2017, from www.knewton.com/infographics/gamification-education/

Gee, J. P. (2005). *Why video games are good for your soul: Pleasure and learning.* Common Ground.

Kilpatrick, J., Swafford, J., & Findell, B. (2001). *Adding it up: Helping children learn mathematics.* National Academy Press.

Knewton. (n.d.). Retrieved March 23, 2019, from www.knewton.com (link doesn't work anymore)

Maccini, P., & Gagnon, J. C. (2000). Best practices for teaching mathematics to secondary students with special needs. *Focus on Exceptional Children, 32,* 1–21.

National Council of Teachers of Mathematics. (1989). *Curriculum and evaluation standards for school mathematics.* Author.

National Council of Teachers of Mathematics. (2000). *Principles and standards for school mathematics.* Author.

Newton, N. (2010). *The Dolch words of math.* Retrieved December 29, 2000, from https://guidedmath.wordpress.com/2010/06/08/the-dolch-words-of-math/

O'Connell, S., & SanGiovanni, J. (2011). *Mastering the basic facts in multiplication and division.* Heinemann.

Parrish, S. (2010). *Number talks: Helping children build mental math and computation strategies, grades K-5.* Math Solutions.

Prensky, M. (2001). Digital natives, digital immigrants part 1. *On the Horizon, 9*(5), 1–6. https://doi.org/10.1108/10748120110424816

Stickney, E. M., Sharp, L. B., & Kenyon, A. S. (2012). Technology-enhanced assessment of math fact automaticity: Patterns of performance for low- and typically achieving students. *Assessment for Effective Intervention, 37*(2), 84–94. https://doi.org/10.1177/1534508411430321

Stille, D. (2017). *Meaningful instruction of basic multiplication facts: Applying constructivist concepts to basic fact acquisition* [Master of Education Program Theses], 111. http://digitalcollections.dordt.edu/med_theses/111

Van de Walle, J. A. (2001). *Elementary and middle school mathematics: Teaching developmentally* (4th ed.). Addison Wesley Longman, Inc.

Van de Walle, J. A. (2007). *Elementary and middle school mathematics: Teaching developmentally*. Pearson/Allyn and Bacon.

Vygotsky, L. S., & Cole, M. (1978). *Mind in society: Development of higher psychological processes*. Harvard University Press.

Ward, R. (2005). Using children's literature to inspire K-8 preservice teachers' future mathematics pedagogy. *The Reading Teacher, 59*(2), 132–143.

Whitin, D. J., & Whitin, P. (2004). *New visions for linking literature and mathematics*. National Council of Teachers of English.

Whitin, D. J., & Wilde, S. (1992). *Read any good math lately? Children's books for mathematical learning, K-6*. Heinemann.

Whitin, D. J., & Wilde, S. (1995). *It's the story that counts: More children's books for mathematical learning, K-6*. Heinemann.

Wilburne, J., & Napoli, M. (2008). Connecting mathematics and literature: An analysis of pre-service elementary student teachers' changing beliefs and knowledge. *IUMST: The Journal, 2*.

Zambo, R. (2005). The power of two: Linking mathematics and literature. *Mathematics Teaching in the Middle School, 10*(8), 394–399.

CHAPTER 2

Modeling Math Facts

Good manipulatives and good education with manipulatives provides students with meaningful material from which students can build, strengthen, and connect powerful representations of mathematical ideas.

(Sarama & Clements, 2016)

Modeling multiplication and division math strategies helps build conceptual understanding. Manipulatives, both physical and virtual, are essential in the teaching and learning of mathematics (see Figures 2.1 and 2.2) (NCSM, 2013; Wenglinsky, 2000). Concrete manipulatives are objects that students can touch, hold, feel, move around, and think with. Digital manipulatives are virtual tools that students can view and move around on the screen to show their thinking. Van de Walle et al. (2013) defines a math tool as

any object, picture, or drawing that represents a concept or onto which the relationship for that concept can be imposed. Manipulatives are physical objects that students and teachers can use to illustrate and discover mathematical concepts, whether made specifically for mathematics (e.g., connecting cubes) or for other purposes (e.g. buttons).

(p. 24)

Manipulatives allow students to see the quantities they are working with and to model the math that they are learning. All brains have visual parts that can be developed and strengthened by visualizing abstract concepts, providing access to all students to learn the concepts. Manipulatives also provide a way for students to talk about their thinking (Hartshorn & Boren, 1990; Ruzic & O'Connell, 2001). Manipulatives can be engaging and inviting. We say inviting because they "invite" students to play around with the math they are learning. Students who work with manipulatives have been found to be more

> In order to develop every student's mathematical proficiency, leaders and teachers must systematically integrate the use of concrete and virtual manipulatives into classroom instruction at all grade levels. (NCSM, 2013)

FIGURE 2.1 Consistently Use Tools

DOI: 10.4324/9781032614229-2

When students are exposed to hands-on learning on a weekly rather than a monthly basis, they prove to be 72% of a grade level ahead in mathematics. Wenglinsky's (2000) (p. 27)

FIGURE 2.2   A Hands-On Approach

interested in math, and students who are more interested have been found to tend to stick with it longer and achieve more (Sutton & Krueger, 2002).

The 2021 What Works Clearinghouse report notes that visual models and manipulatives are among the research-based recommendations to develop understanding. More specifically, when manipulatives are included in the instructional sequence, the concrete–representational–abstract (CRA) approach is supported. In this approach, students use concrete manipulatives to build understanding and see patterns, structure, and number relationships. With the concrete manipulatives built, students can then begin recording their work with manipulatives using pictorial mathematical drawings and sketches to represent their thinking. The concrete and pictorial representations can also be recorded abstractly using equations. Rather than progressing from concrete to pictorial and then abstract, the more we have our students connect these representations simultaneously, the better. Our friend Christina Tondevold says that when we represent our thinking using concrete, pictorial, and abstract approaches, then that's the "sweet spot" (see Figure 2.3).

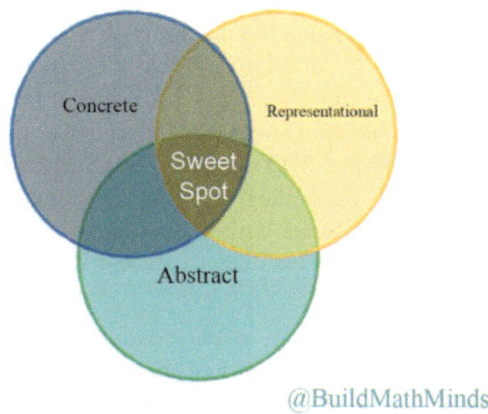

FIGURE 2.3   Concrete Representational Abstract Sweet Spot

STRATEGIES AND MODELS

There is a difference between strategies and models (see Figure 2.4). Strategies are what we do with numbers, such as skip counting, finding partial products to help find products, or doubling and halving. Models are how we make this thinking visible and can be concrete, pictorial, or abstract. Oftentimes, charts will use the term "strategies," but they are actually showing models. We want to be sure that all teachers and students are clear about the difference between the two.

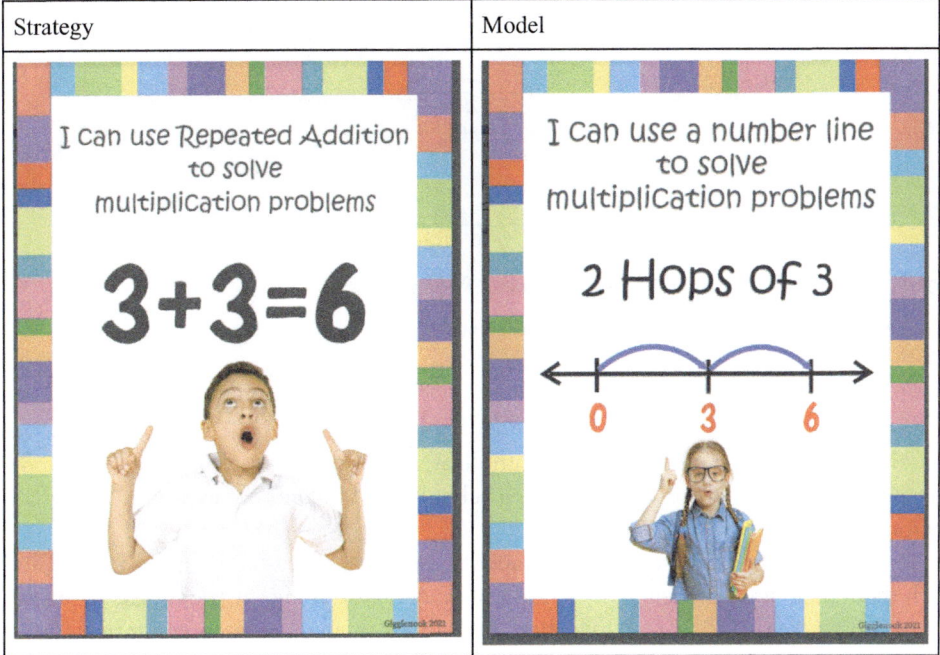

FIGURE 2.4   Strategies and Models

Some examples of strategies when solving multiplication problems include counting each item in each of the equal groups (counting phase of reasoning), skip counting (additive reasoning), and using facts that are known to determine facts that are not yet known (multiplicative reasoning). Here is a graphic by Pam Harris showing the K–12 math journey of our students (see Figure 2.5).

Pam Harris refers to this graphic as the development of mathematical reasoning. (www. mathisfigureoutable.com). The progression outlined in the graphic is impactful for us, as it helps to determine where our students are on the reasoning continuum when they share their thinking with us. From birth to around kindergarten age, students are expected to be in a counting phase of reasoning. In grades 1 and 2, their reasoning transitions into additive

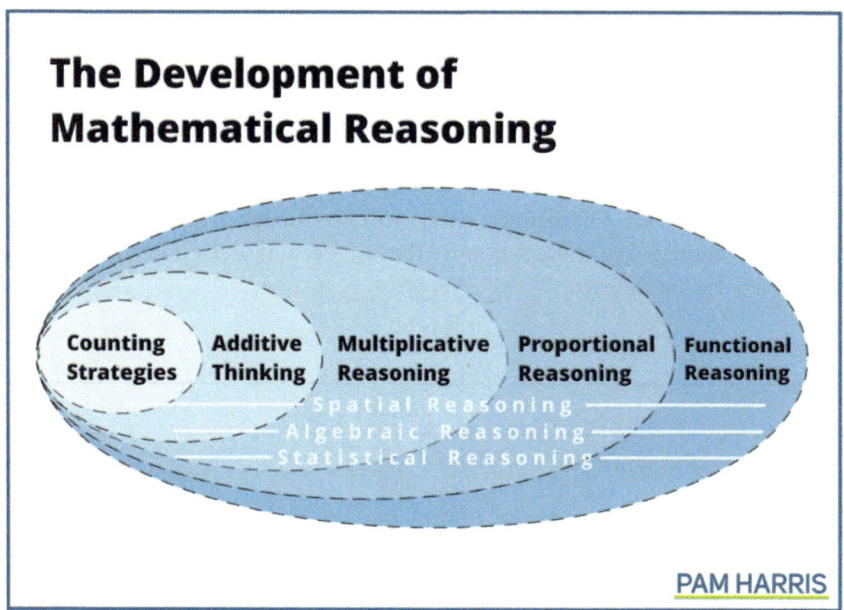

FIGURE 2.5 The Development of Mathematical Reasoning

(www.mathisfigureoutable.com This work is licensed under a Creative Commons Attribution-NoDerivatives 4.0 International License)

thinking. In grades 3–5, they explore multiplicative reasoning in preparation for proportional reasoning in grades 6–8. Finally, high-school-aged students are exploring functional reasoning.

Too often, though, our students are caught in a phase of reasoning that is not aligned with their grade-level content. For example, if students are using the traditional algorithm for two-digit by two-digit multiplication, very often they are skip counting to find the multiplication math facts within the computation work. While they are technically solving a multiplication problem, their skip counting approach reflects additive reasoning, not the multiplicative reasoning of the multi-digit multiplication content. Using manipulatives to cultivate multiplicative reasoning, beginning with basic facts, fosters number sense and builds understanding of number relationships. This conceptual foundation helps to transition from additive to multiplicative reasoning. This approach not only helps students to develop mastery of their basic facts but also builds a foundation of flexible thinking that will be applied to all sets of numbers they will encounter on their math journeys.

It may be helpful at this point to explore one strategy through various models and also to explore one model to show various strategies. Let's take the example of 4 × 6. One strategy we can use is to think of multiplying by 4 as the same as doubling the other factor and then doubling again. This is the heart of the distributive property of multiplication and will be able to follow our students through their math journeys with the various sets of numbers they will encounter. This strategy can be made visible in concrete, pictorial, and abstract ways (see Figure 2.6).

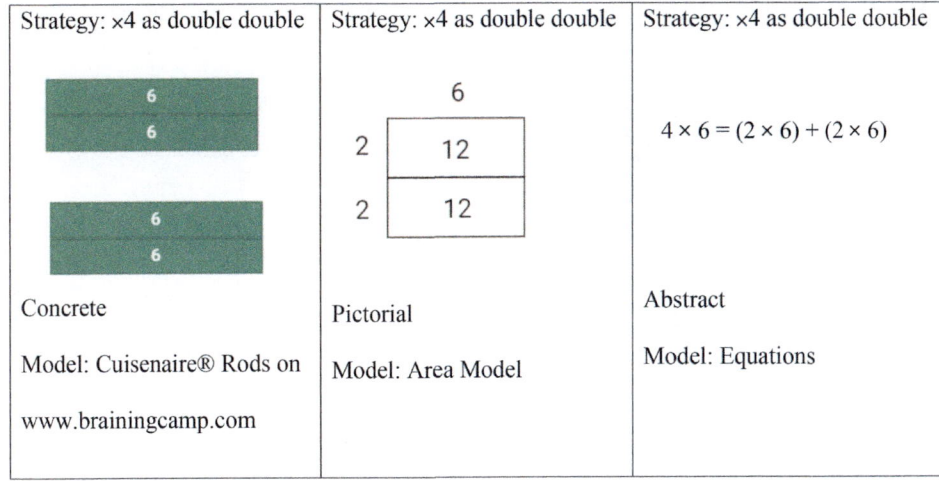

Strategy: ×4 as double double	Strategy: ×4 as double double	Strategy: ×4 as double double
		$4 \times 6 = (2 \times 6) + (2 \times 6)$
Concrete	Pictorial	Abstract
Model: Cuisenaire® Rods on www.brainingcamp.com	Model: Area Model	Model: Equations

FIGURE 2.6   Example of CRA Progression of 4 × 6

We can also use one tool, such as the rekenrek, to model a variety of strategies. When students have developed flexibility through the use of various strategies, they will then be able to pick a strategy that works most efficiently for them given the numbers in the problem. Here are few strategies that can be used to solve 4 × 6, all using the rekenrek to model the strategies (see Figure 2.7).

Strategy: ×4 as double double	Strategy: Decompose 6 into 5 groups and 1 more group	Strategy: Decompose 6 into two groups of 3
Model: Rekenrek on www.brainingcamp.com	**Model:** Rekenrek on www.brainingcamp.com	**Model:** Rekenrek on www.brainingcamp.com

FIGURE 2.7   Models

Here is a sample of some anchor charts to illustrate the difference (see Figure 2.8).

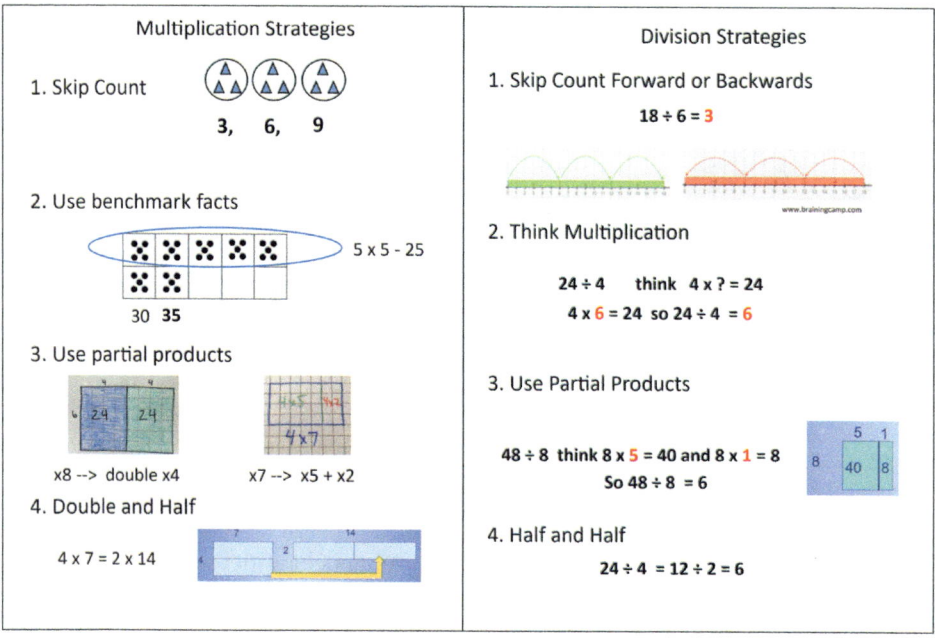

FIGURE 2.8 Strategy Posters

THE CLASSROOM AS A TOOLKIT

First, your classroom can be your largest toolkit (see www.pinterest.com/drnicki7/math-toolkits-classroom-as-toolkit). It should have several of the tools that are used in large, student-size form. For example, there should be number lines that students can walk on and see on the walls. There should be a large hundreds grids at the board for students to work on. Particularly when students are developing their foundational understanding of what multiplication and division mean, teachers should have large circles on the floor made by duct tape or hula hoops so that students can act out multiplication and division problems kinesthetically (see Figures 2.9 and 2.10). Acting out a problem is different from modeling it with manipulatives and it is so important that students have an opportunity to do both. There should also be posters up that support concepts and vocabulary (see Figures 2.11 and 2.12).

FIGURE 2.9 Acting Out Situations with Objects and Hula Hoops

FIGURE 2.10 * Acting Out Kinesthetically Using Hula Hoops

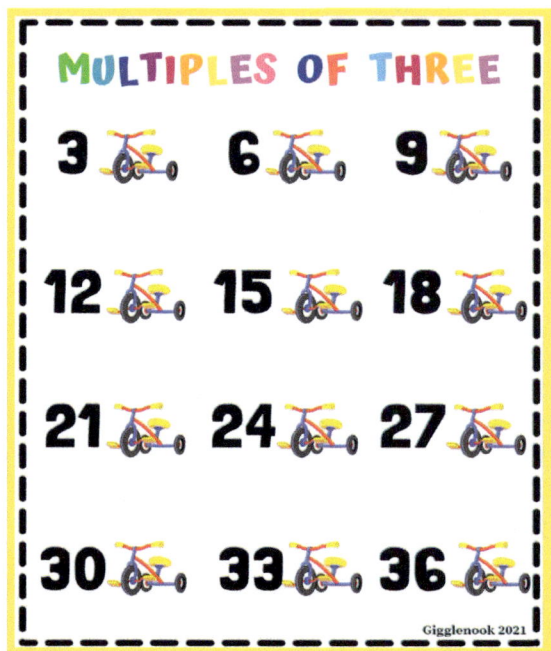

FIGURE 2.11 Multiples of 3

FIGURE 2.12   Vocabulary

INDIVIDUAL STUDENT TOOLKITS

We believe that individual toolkits are an important part of math class and that students need them within arm's reach so that they can turn to and use them at any moment (rather than stored in buckets on a bookshelf that they would need to walk over to get if they "need" it) (see Figure 2.13–2.15). They don't have to be specific to each student,

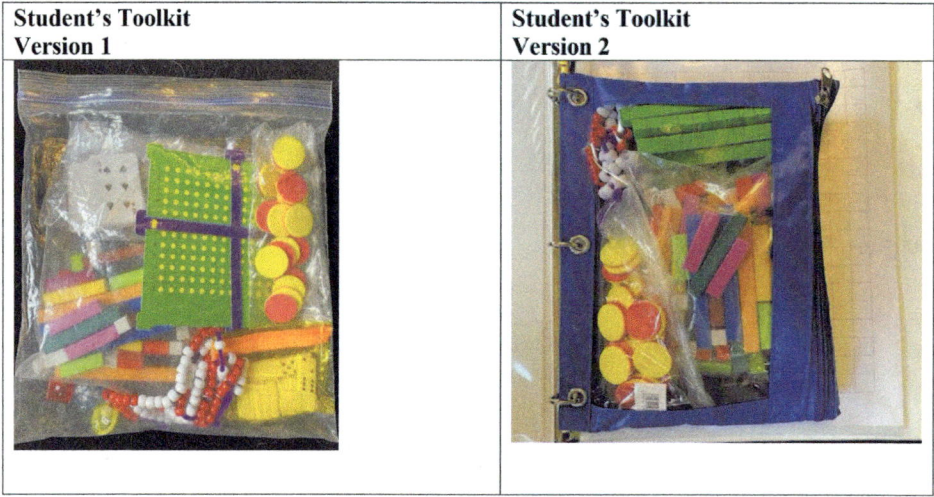

FIGURE 2.13 Toolkits

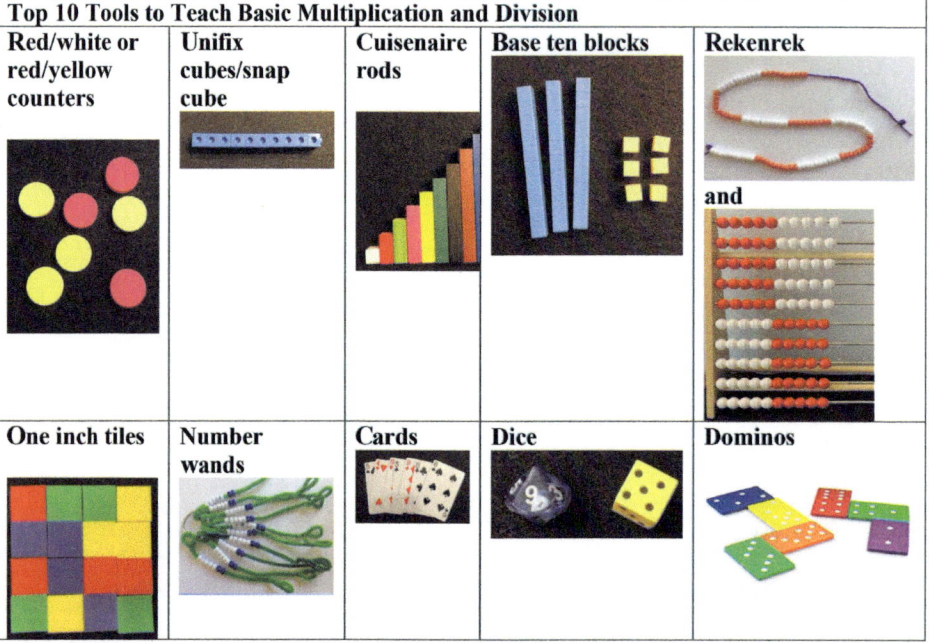

FIGURE 2.14 Tools

although they could be. However, there should be a place where the templates and the tools are centrally located so that when the teacher pulls up the number line on the smartboard, students can open up to their laminated number line and either act out the same problem with manipulatives or draw it out. The items in the toolkits at the beginning of the year would be manipulatives they are familiar with from the previous year. Over the course of the school year, you would add to the toolkits gradually as you introduce new tools and templates to the students. Students absolutely need toolkits because **math is NOT a spectator sport,** and they need to play the game, not watch it!

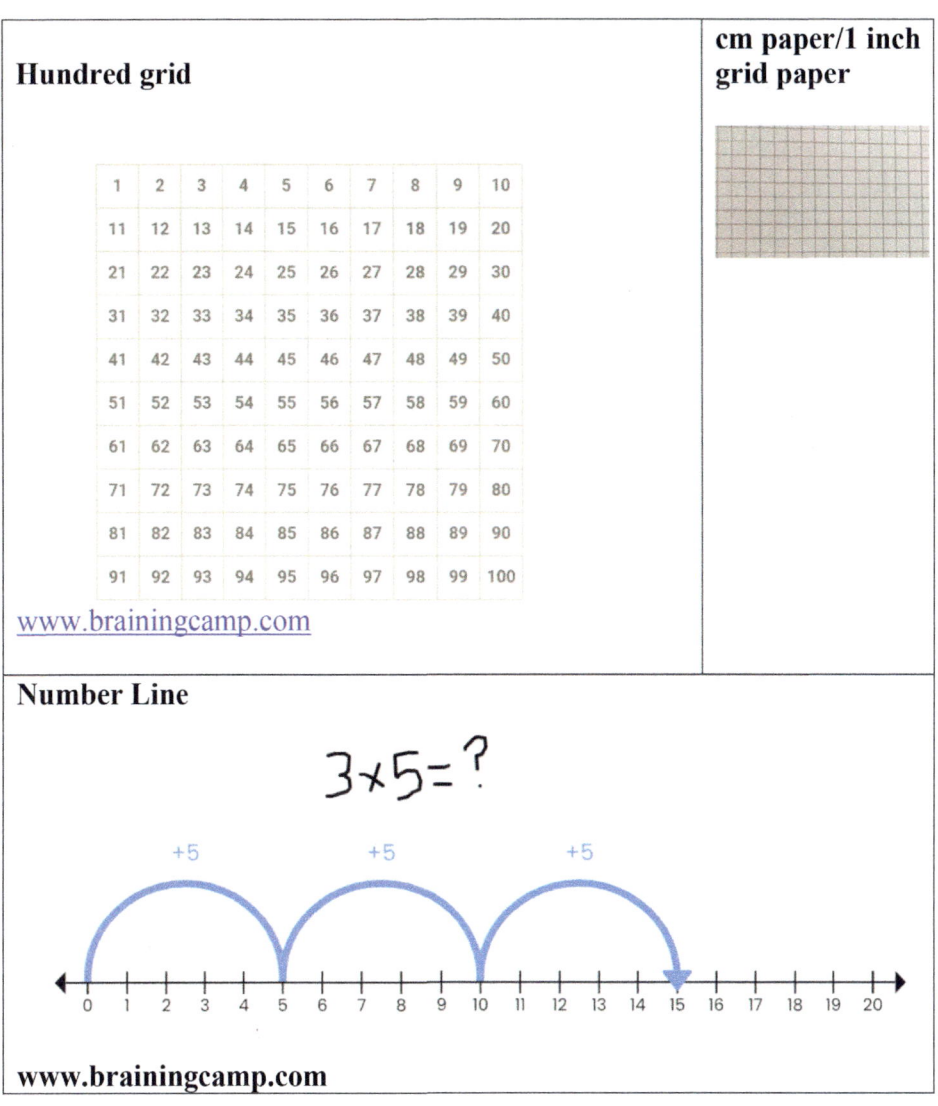

FIGURE 2.15 Templates

TEACHER TOOLKITS

The teacher should have a toolkit easily accessible that mirrors the student toolkits. In this way, the teacher isn't running around trying to get the basic tools every time to model something. One of the additional things that the teacher can have in their toolkit is magnetized manipulatives. They can be bought or made. To make them, you simply have to buy magnet tape and glue it onto the manipulatives.

DIGITAL TOOLKITS

There are so many wonderful digital tools out there now. We are amazed at how many there are and the different types of things that you can do with them (see Figure 2.16). In the following figure are some of our favorites:

Virtual Rekenrek	Virtual Number Lines Open and Marked Lines	Virtual Number Bead
Braining Camp 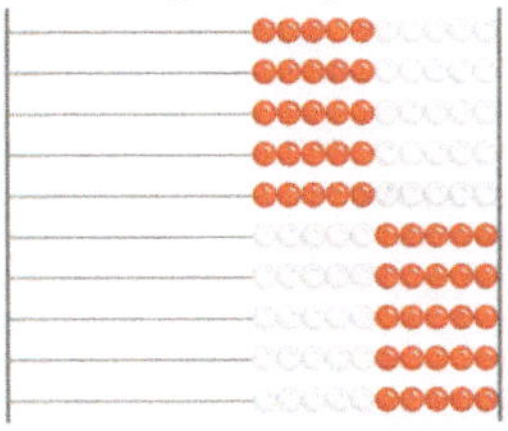	**Braining Camp** 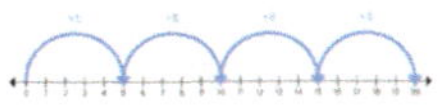	 www.brainingcamp.com
Virtual Cuisenaire Rods	**Virtual Hundreds Grid**	**Virtual Money**
https://www.brainingcamp.com 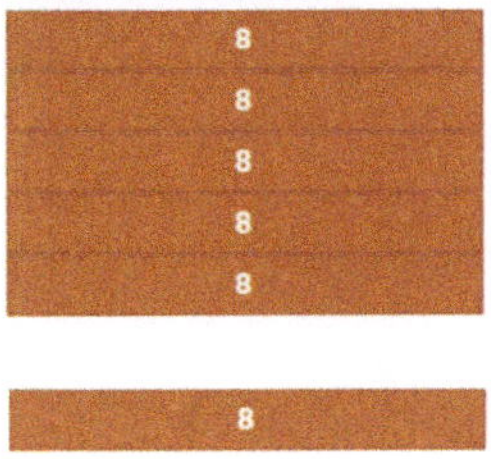	**Braining Camp** 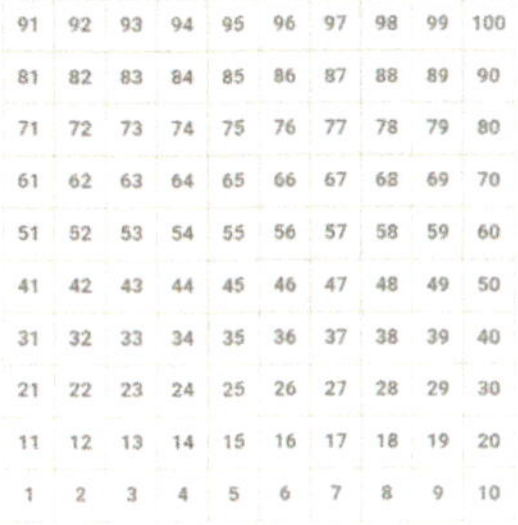	**Braining Camp** 

FIGURE 2.16 Digital Tools

ANCHOR CHARTS

It is important that the teacher makes anchor charts with the students so that they can discuss, explain, and understand the different strategies. Teachers can put a QR code on these charts, take a picture, and send them home to parents so that parents can watch videos of the strategies in action. These videos can be made by the teacher or the students, or pulled from YouTube, TeacherTube, or SchoolTube (see Figure 2.11). Make sure there is a place in the classroom where these can stay up for most if not all of the year. As Dr. Nicki always says, "The classroom walls tell us what is valued in the classroom." Moreover, students should have an opportunity to make their own charts. Students can keep their charts in their math journal under the math strategy section.

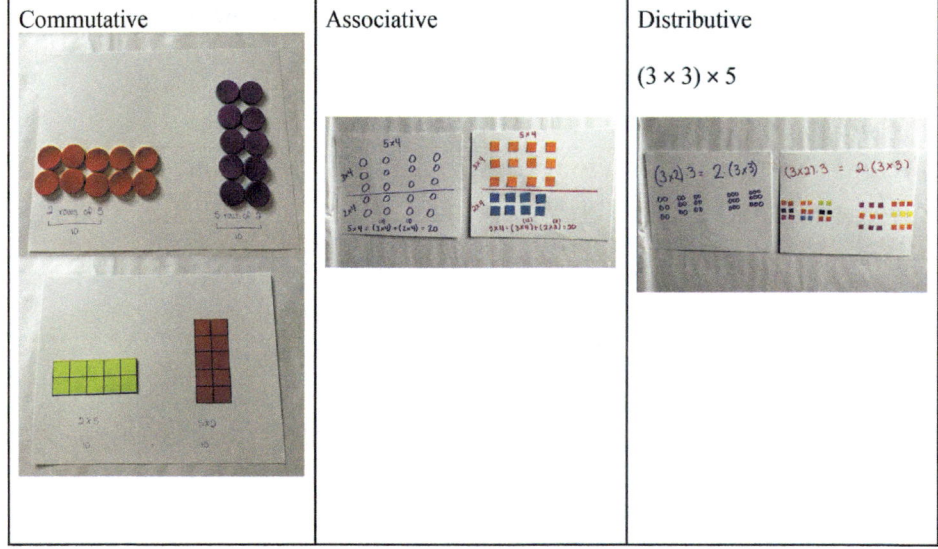

FIGURE 2.17   Anchor Charts

KEY POINTS

- Modeling
- Cycle of engagement—concrete, pictorial, and abstract
- Concrete tools
- Digital tools
- The classroom as a toolkit
- Individual student toolkits
- Teacher toolkits

SUMMARY

Modeling math facts is an essential part of developing a conceptual understanding of not only how numbers relate to each other but also the operations of multiplication and division. We model our thinking concretely and virtually with tools, pictorially using area models and number lines, and abstractly with equations. We want students to go through the cycle of engagement—concrete, pictorial, and abstract—and to connect these representations as much as possible. Students should have their own toolkits with templates that they can write on and tools that they have within arm's reach. Teachers should have their own toolkits as well so that they can have easy access to the templates and tools that they need during mini lessons and guided math lessons. This is not to say that everything is in a toolkit. Rather, this is to say that there is a specific set of everyday tools that students have instant access to when they need it.

REFLECTION QUESTIONS

1. In what ways are you modeling facts currently?
2. Are you using your classroom as a toolkit with big models of the tools?
3. What are some takeaways you have from this chapter?

CALL TO ACTION

 1. Share your favorite "Aha!" moment from this chapter on social media to help spread the movement! #FDJH

 2. Take a photo of different ways that you are modeling math facts in your classroom and share it on social media to encourage other teachers to do it too! #FDJH

REFERENCES

Harris, P. (2022). Retrieved December 29, from www.mathisfigureoutable.com/

Hartshorn, R., & Boren, S. (1990). *Experiential learning of mathematics: Using manipulatives.* ERIC Clearinghouse on Rural Education and Small Schools.

Institute of Educational Science. (2021). *Doing what works clearinghouse.* Retrieved December 29, 2022, from https://ies.ed.gov/ncee/wwc/practiceguide/2

National Council of Supervisors of Mathematics. (2013). *Improving student achievement in mathematics by using manipulatives with classroom instruction.* NCSM Position Paper.

Ruzic, R., & O'Connell, K. (2001). *Manipulatives.* National Center on Accessing the General Curriculum. www.cast.org/ncac/index.cfm?i=1666

Sarama, J., & Clements, D. H. (2016). Physical and virtual manipulatives: What is "concrete"? In P. S. Moyer-Packenham (Ed.), *International perspectives on teaching and learning mathematics with virtual manipulatives* (pp. 71–93). Springer International Publishing.

Sutton, J., & Krueger, A. (2002). *What we know about mathematics teaching and learning.* Mid-continent Research for Education and Learning.

Van de Walle, J. A., Karp, K. S., & Bay Williams, J. M. (2013). *Elementary and middle school mathematics: Teaching developmentally* (8th ed.). Pearson.

Wenglinsky, H. (2000). *How teaching matters: Bringing the classroom back into discussions of teacher quality.* Educational Testing Service.

CHAPTER 3

Exploring and Learning Multiplication Facts

When parents prepare to enroll their children in school, they likely anticipate that one thing they will have to help them with is learning their basic multiplication facts. For many, their own experience likely involved rote memorization, games like around the world, drilling flashcards, and being subjected to timed tests. We estimate that 80% of the parents, teachers, and Uber drivers that we encounter freely share their negative math experiences using words such as tears, memorization, fear, stress, and anxiety when they discover that we are math consultants. When asked when they first began feeling that way, most share stories of taking timed tests and feeling defeated when they weren't speedy at math. They share tales of the displays that were intended to be walls of fame but too often became walls of shame. Their hot air balloons never rose to the sky, their race cars never moved on the tracks, and they didn't earn more than a scoop of ice cream on their sundaes. The damage and impact of these practices is far reaching and becomes a part of the identity of those who experience them. Many adults who declare themselves as "not math people" track the genesis of this belief back to timed tests, being a slower processor, and not being able to produce answers instantly and on demand.

It comes as a surprise to many with that mindset that fluency has nothing to do with speed. Instead, fluency is defined as flexibility, efficiency, accuracy, and appropriate strategy selection when computing with numbers (National Research Council, Mathematics Learning Study Committee, 2001; National Council of Teachers of Mathematics, 2014; Bay-Williams & SanGiovanni, 2021). With this definition in mind, it is clear that timed tests are not a true assessment of fluency. In addition, the unspoken message that speed equates to ability has adverse effects on students' math dispositions and fosters anxiety with math (Boaler et al., 2015). While drilling facts and focusing on speed are damaging to math dispositions, they also do nothing to build flexibility. Flexibility is an important outcome that supports students in eventually multiplying larger numbers and solving unfamiliar facts. Using a rhyme of "eight and eight fell on the floor . . . picked them up is 64" only helps students when they encounter 8 × 8 and doesn't even help with that when forgotten.

Yet if we take the time to develop number relationships and allow students to discover that the 8's facts are double the 4's facts, students are set up for success to multiply anything times 8. When we explore the relationship of the 9's as one group less than the 10's, students are empowered to solve anything times 9. These experiences serve to cultivate positive math dispositions because students are *thinking* rather than

DOI: 10.4324/9781032614229-3

memorizing . . . an important distinction (Fosnot & Dolk, 2001). Although students may not know an answer instantly, they have confidence to use what they already know to deduce it. Slowing down to build a foundation for *all* aspects of fluency, and fostering a classroom environment that cultivates opportunities for students to grow in each area, creates a framework for thinking that will follow students all the way through their math journeys.

INTRODUCING THE FACTS

You may notice that the order of facts we suggest does not follow the sequential order of the factors. Indeed, teaching the facts in order of factor size is a practice that Kling and Bay-Williams (2021) name as one of the eight unproductive practices for building fact fluency. Instead, we want to explore strategies based on a number relationships (Bay-Williams & SanGiovanni, 2021; Baroody, 2006; Brendefur et al., 2015). While the exact order differs among various scholars, they are all in agreement that the foundational facts of ×0, ×1, ×10, ×5, and ×2 must be mastered to help facilitate the process of deriving larger factors.

The heart of the grade 3–5 multiplication journey is to facilitate students' progression from additive reasoning to multiplicative reasoning. In this chapter, we will provide a plethora of ideas to explore number relationships through concrete, pictorial, and abstract representations so that every student may develop a foundation of flexibility, efficiency, accuracy, and strategic competence with basic multiplication math facts, while setting a foundation for the multiplication concepts yet to come.

LEARNING TRAJECTORIES INVOLVING MULTIPLICATION

As we facilitate the math journeys of our students, it is helpful to understand the research behind the learning trajectories. This will help to ensure that we offer purposeful practice with certain sets of facts and that we leverage targeted, small-group instruction to facilitate the progression from counting to additive reasoning and then to multiplicative reasoning. The seminal research from Sherin and Fuson (2005) is a must-read. We particularly love their chart of the progression on page 358. Through our exploration of multiplication using concrete, pictorial, and abstract representations, we want to help students move into multiplicative reasoning by using multiplicative relationships or finding groups of groups within the total. For example, students can explore the doubling relationships between ×2, ×4, and ×8; examine the halving relationship between ×10 and ×5; and use facts they know to determine facts they don't yet. For example, if a student was struggling with 4×7, they could think about it as $(4 \times 2) + (4 \times 5)$, two facts that they have already mastered. Understanding these relationships helps students to reason strategically rather than falling back to the additive reasoning of counting. Building multiplicative reasoning is empowering because the same strategies used with basic facts can be applied to multi-digit numbers, decimals, and fractions.

PROPERTIES OF MULTIPLICATION

Before becoming truly fluent with basic multiplication facts, students should explore activities that allow them to develop conceptual understanding of the operation of multiplication as well as the properties of multiplication. Math standards specifically state that students need to learn how to calculate with the four operations using *strategies based on place value, properties, and the relationships between the operations*. Equipped with this understanding, students may leverage the properties to more flexibly and efficiently arrive at their products. Let's explore the following properties of multiplication: commutative, associative, and distributive.

COMMUTATIVE PROPERTY OF MULTIPLICATION

One of the first properties students encounter is the commutative property of addition as they explore that 2 + 6 results in the same sum as 6 + 2. When students are in a counting phase of reasoning, this property is powerful because rather than counting on 6 from the 2, they can understand that starting at the 6 and counting on the 2 will provide an equivalent answer with more efficiency and accuracy. Beginning grade 3, this understanding of the commutative property of addition is extended as students explore the commutative property of multiplication. In the U.S., the convention is to read "10 × 8" as "10 groups of 8." (This is not a hard and fast math rule, though, as other countries read it as "10 eight times.") As students begin to explore multiplication, they often skip count to find the products. In this example they might skip count by 8's ten times. However, if we have explored the commutative property of multiplication, they will instead understand that they can change the order of the factors and perhaps think of the problem in a more efficient way to find the product. The "10 groups of 8" can become "8 groups of 10", which many know quickly as 80. As always, exploring concepts in concrete and pictorial ways is helpful before we jump to the abstract expressions. While there are many ways to represent multiplication concretely, one of our favorite manipulatives is Cuisenaire 4 rods because they model how the equivalent expressions have the same area (see Figure 3.1). This is great for students to discover.

FIGURE 3.1 Commutative Property Shown with Cuisenaire® Rods

This area model of 3 of the 4 rods is the same quantity as the model of 4 of the 3 rods. So, 3 × 4 = 4 × 3.

ASSOCIATIVE PROPERTY OF MULTIPLICATION

The associative property allows us to strategically join together addends or factors in ways that allow us to find our sums or products more efficiently. For example, with addition, if students are asked 4 + (6 + 7), they can think about associating, or grouping, the 4 and 6 together so that the expression would read (4 + 6) + 7, becoming simply 10 + 7. Similarly, with multiplication, students can multiply two of the three factors by associating them to more efficiently determine their products. For example, 4 × (5 × 7) = 4 × 35 can be renamed as (4 × 5) × 7 = 20 × 7, which is more accessible for many more of our students as an extension of the basic fact of 2 × 7.

The associative property is also at the heart of strategies when some addends or factors are decomposed first and then reassociated. For example, 98 + 57 can look pretty scary to some students, but if we think about decomposing 57 into 2 and 55, then 98 + 57 can be renamed as 98 + (2 + 55) and then the addends can be associated to think (98 + 2) + 55, thus 100 + 55. Similarly, with multiplication, when students are asked 5 × 8, they may think about breaking apart the 8 into 2 × 4 so that the expression can be renamed as 5 × (2 × 4) and associated to (5 × 2) × 4, thus 10 × 4. This particular example associating the factor of 2 is at the heart of what is more commonly referred to as the double and halving strategy. This strategy may also be extended to triples and thirds, quadruples and fourths, and so on, well beyond the scope and quantity of basic facts.

DISTRIBUTIVE PROPERTY OF MULTIPLICATION

The distributive property of multiplication allows students to be flexible in breaking apart one or both of the factors and distributing one factor to the decomposed parts. This begins with basic facts and is then extended to larger numbers as well. One powerful example with basic facts is the ×7 facts. In our experience administering Math Running Records (www.mathrunningrecords.com), when students are asked which expressions look tricky, they inevitably mention the expressions that include a 7. If we consider, however, decomposing 7 into 5 and 2 and then distributing the 2 and 5 to whatever other factor is being multiplied, we are using the foundational facts of ×2 and ×5 to determine the total of ×7. For example, 8 × 7 = (8 × 5) + (8 × 2), which doesn't seem so scary as first thought. For many students, it is important to make the decompositions visible and show the equivalent area, as seen in Figure 3.2.

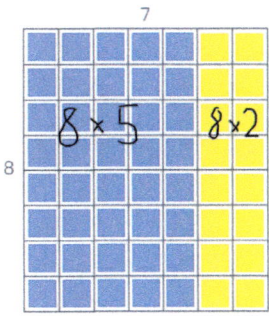

FIGURE 3.2   Colored Tile Array of 8 × 7

This property extends to double-digit multiplication, when students may decompose a number by place value and multiply each piece by the other factor. For example, $26 \times 4 = (20 \times 4) + (6 \times 4)$. Students continue to leverage this strategy as they encounter multiplication involving multi-digit numbers, mixed numbers, decimals, and even polynomials.

Students should have the opportunity to explore all of the properties and number relationships through concrete, pictorial, and abstract representations during whole-group, small-group, and workstation learning times (see Figure 3.3 and 3.4).

TOOLS, TEMPLATES, ACTIVITIES, AND GAMES

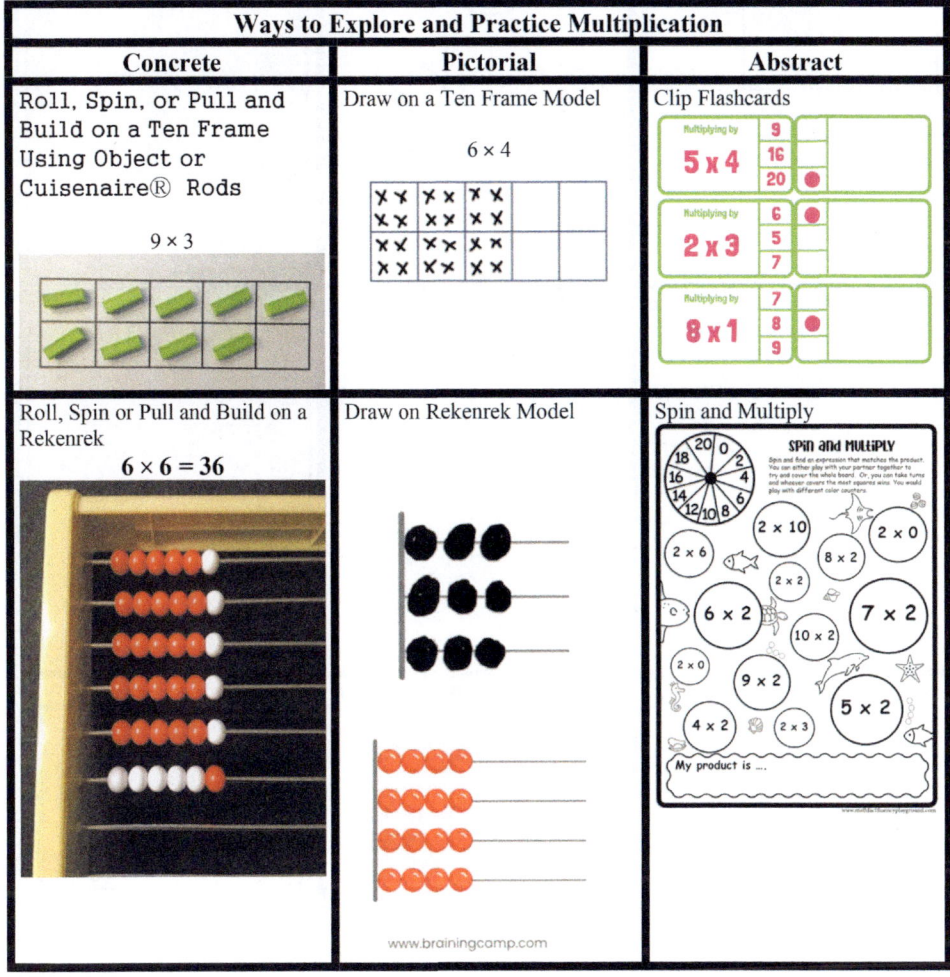

Ways to Explore and Practice Multiplication		
Concrete	**Pictorial**	**Abstract**
Roll, Spin, or Pull and Build on a Ten Frame Using Object or Cuisenaire® Rods 9 × 3	Draw on a Ten Frame Model 6 × 4	Clip Flashcards
Roll, Spin or Pull and Build on a Rekenrek 6 × 6 = 36	Draw on Rekenrek Model www.brainingcamp.com	Spin and Multiply

FIGURE 3.3 Explorations

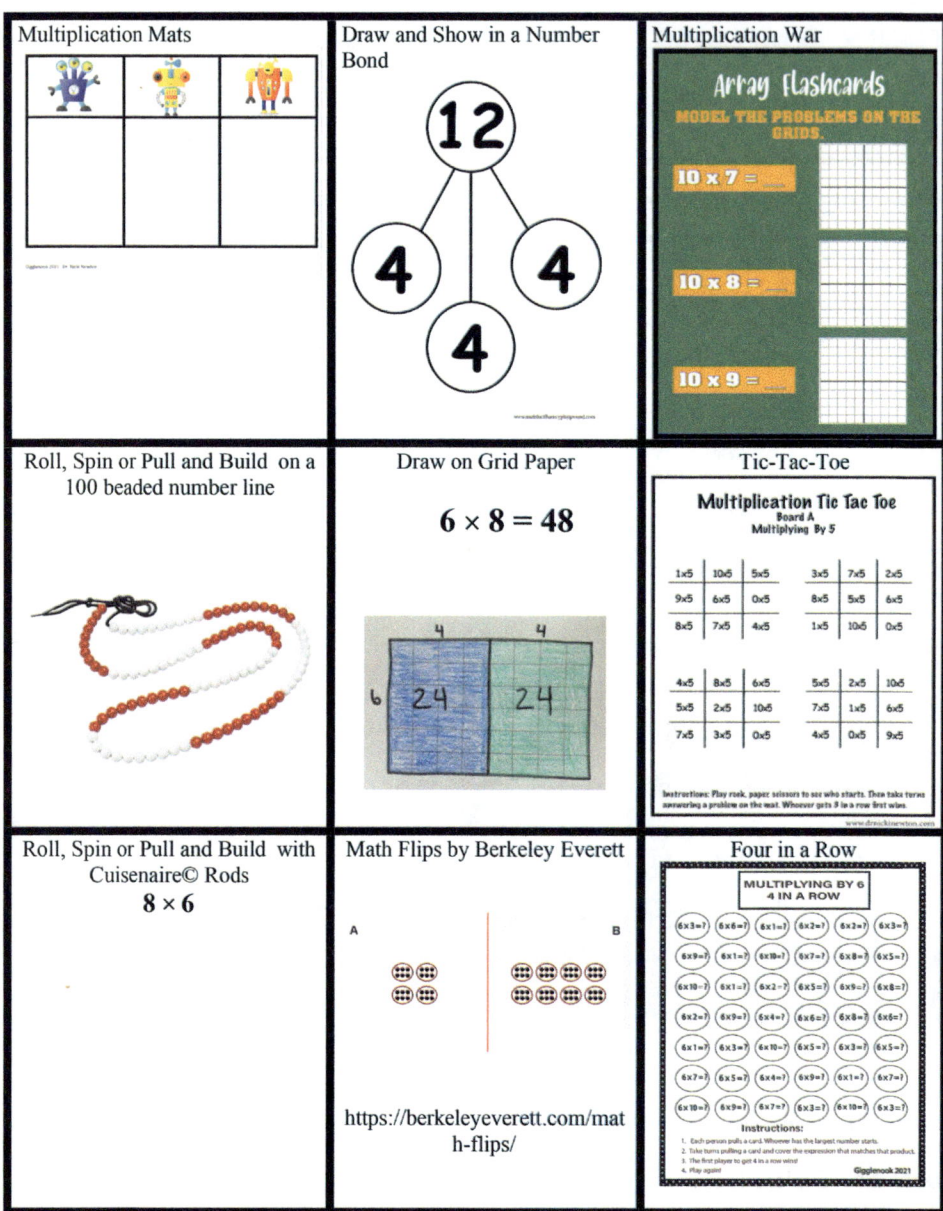

Multiplication Mats	Draw and Show in a Number Bond	Multiplication War
	12 = **4** + **4** + **4**	Array Flashcards — MODEL THE PROBLEMS ON THE GRIDS. 10 x 7 = , 10 x 8 = , 10 x 9 =
Roll, Spin or Pull and Build on a 100 beaded number line	Draw on Grid Paper $6 \times 8 = 48$	Tic-Tac-Toe — Multiplication Tic Tac Toe, Board A, Multiplying By 5
Roll, Spin or Pull and Build with Cuisenaire© Rods 8×6	Math Flips by Berkeley Everett A B https://berkeleyeverett.com/math-flips/	Four in a Row — MULTIPLYING BY 6, 4 IN A ROW

FIGURE 3.3 (Continued)

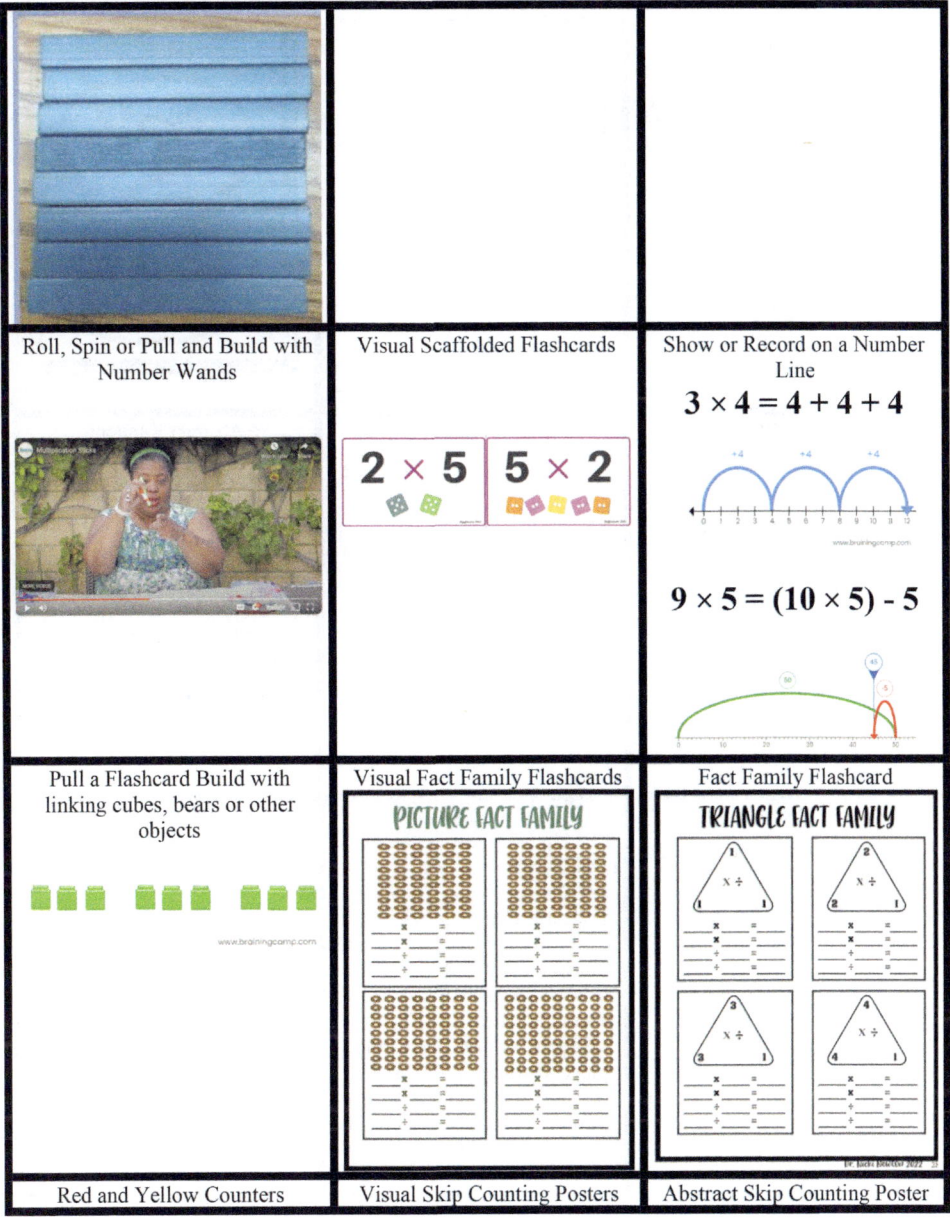

Roll, Spin or Pull and Build with Number Wands	Visual Scaffolded Flashcards	Show or Record on a Number Line $$3 \times 4 = 4 + 4 + 4$$ $$9 \times 5 = (10 \times 5) - 5$$
Pull a Flashcard Build with linking cubes, bears or other objects	Visual Fact Family Flashcards	Fact Family Flashcard
Red and Yellow Counters	Visual Skip Counting Posters	Abstract Skip Counting Poster

FIGURE 3.3 (Continued)

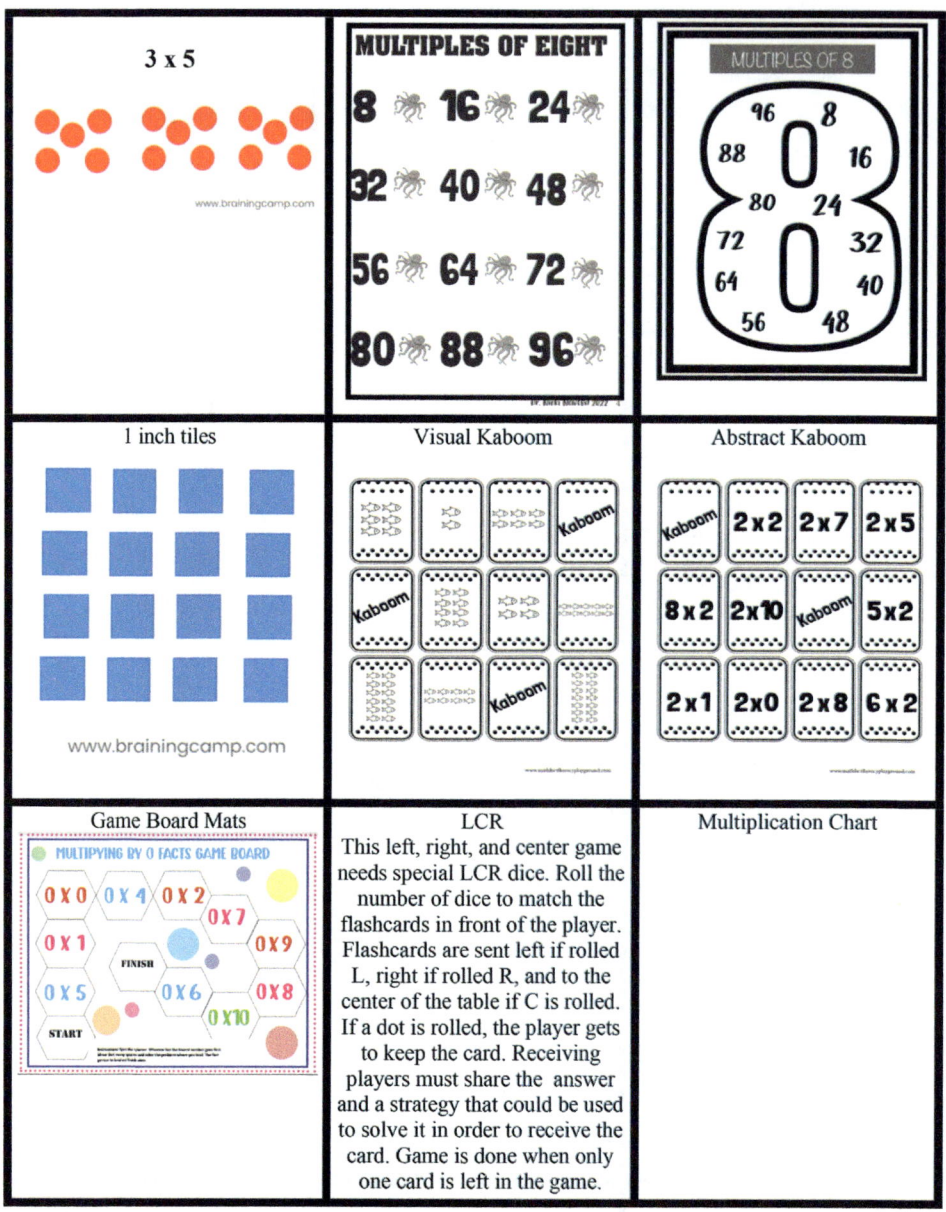

FIGURE 3.3 (Continued)

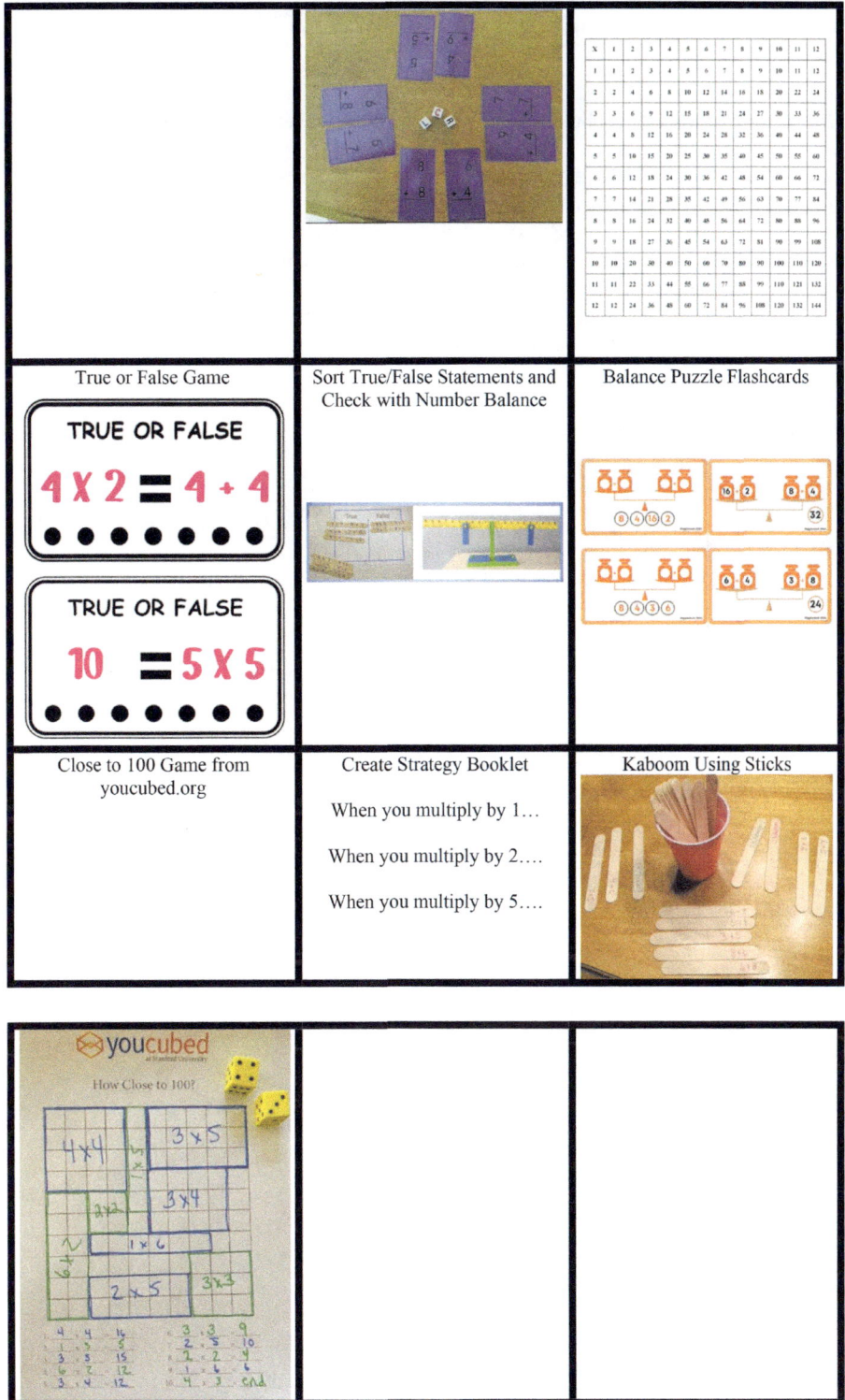

FIGURE 3.3 (Continued)

ONLINE GAMES AND APPS FOR ALL FACTS

Multiplication by Heart - Fabulous set of visual flashcards available both in print available on Amazon as well as online here: https://mathigon.org/multiply

Stick and Split App – This app has students explore the factors of gradually larger products using rods similar to Cuisenaire® Rods. https://www.maypoleeducation.com

Kakooma - This is a game where students are given several numbers and they need to recognize a fact family of two factors and their product. https://tangmath.com/kakooma

Fast Facts App (Don't let the name fool you. You know we don't like timed activities. This app offers visuals to support conceptual understanding.) https://apps.apple.com/us/app/fast-facts-math/id506232953

FIGURE 3.4 Online Games

EXPLORING AND LEARNING ×1 AND ×0 FACTS

Multiplication must be understood conceptually as equal groups and arrays. Once this established, we can begin with ×1 and ×0 facts, which are properties of multiplication. We want students to know not only what the products are but also *why* they are what they are. In our experience interviewing students about their thinking using Dr. Nicki Newton's Math Running Records, students often share that "anything times 0 is 0." We always follow up with, "Do you know why?" Too often, students will say that their teacher told them that "anything times 0 is 0." Sometimes students confuse multiplying by 0 and 1 and say that $7 \times 0 = 7$ rather than 0. Believe it or not, the concept of zero is super tricky for some students! Since we want to promote thinking and understanding, rather than just giving rules, we can make the learning experience interactive by exploring situations with our students. Through thinking and reasoning about familiar contexts, students can develop the understanding that zero groups of a certain size will always result in an amount of zero, as will any number of groups that each have zero objects in them. In the same way, students can discover that when multiplying by 1, the product will always be the other factor because we have either one group of that size or we have that number of groups with 1 in each. We have found it helpful to explore both ×0 and ×1 facts with students when we are working with them in small groups to ensure that students are truly conceptualizing the ideas.

WHOLE-CLASS ACTIVITIES

What's the Story?

Students look at a picture in a slide and decide what the story is. Their task is to come up with an expression that matches the picture. They should be encouraged to describe the picture as well (see Figure 3.5).

The teacher shows the picture and asks the students to tell a story with an equation. Marta says the bakery had 1 box with 4 cupcakes in the box. 1 group of 4 is 4. 1×4 is 4.	

The teacher shows the picture and asks the students to tell a story with an equation. Carl says there is 1 plate with nothing on it. 1 group of 0 is 0. $1 \times 0 = 0$.	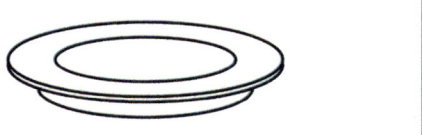

The teacher shows the picture and asks the students to tell a story with an equation. Jamal says the bakery had 5 boxes with 1 cupcake in each box. 5 groups of 1 is 5. 5×1 is 5.	

FIGURE 3.5 What's the Story?

WHOLE-CLASS MINI LESSON: MULTIPLYING BY 0 AND 1

The teacher gives students story mats to act out the story (see Figure 3.6).

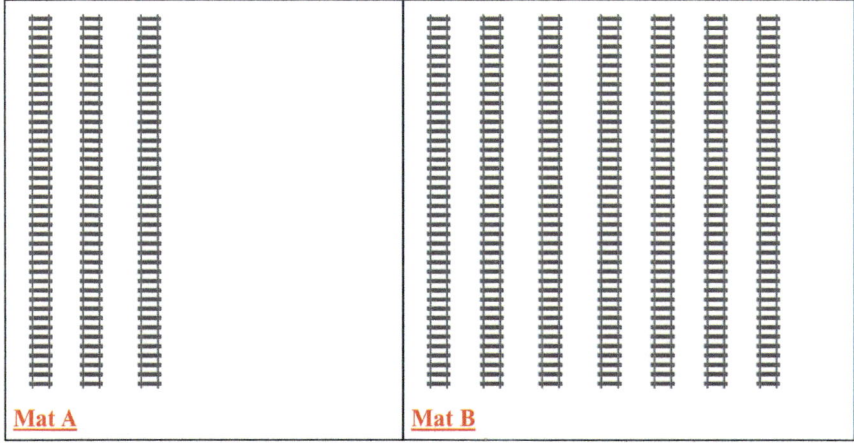

FIGURE 3.6 Story Mats

Introduction: *(Teacher ensures each student has 10 linking cubes and a mat.)*

Teacher: Today we are going to talk about multiplying by 0 and 1.

Addison: That's easy! My mom told me that when we multiply anything times 0, the answer is always 0, and that when you multiply by 1, it is always the other number.

Teacher: That's true, but can you explain why that's true?

Addison: Not really.

Teacher: Let's explore a little bit and I'll come back to see if you can explain why those statements are true. I'm going to tell a story and I'd like you all to use your linking cubes to model the story. At a train station first thing in the morning, there were three tracks and there was a train on each track. Let's pretend that your linking cubes are the trains. Let's model this situation using your cubes.
(Students then use the cubes to model the story situation by placing one cube in three separate spots on their mats.

Teacher: Will someone explain what you did?

Joe: The story said that there were 3 trains on the track. So now we have 3 groups of 1 and that makes 3.

Teacher: Thank you, Joe. Let's all clear our mats. Who wants to tell a story?

Meghan: There was 1 track. There were 2 trains on it. So that is 1 track with 2 trains. So that is 1 group of 2 trains.

Mary: I agree with Meghan. Since there are 2 trains in the station but they are on 1 track.

Teacher: Who wants to tell another story?

Chris: There were 7 tracks and there was 1 train on each. So that is 7 groups of 1 and that makes 7.

Teacher: So, what do we know about multiplying by 0 and 1?

Addison: I think I understand it now. If you have no groups of some size or many groups each with nothing in them, the total amount will always be zero.

Joe: And when we are multiplying by 1, we either have 1 group of a certain size or we have a number of groups each with 1 in them. So, the answer will always be the amount in the number we are multiplying by 1.

Teacher: Exactly! It's all about how many groups we have and how many are in each of those groups.

 ## SPOTLIGHT ACTIVITY

Herding Game

For this activity, you can play the herding game (great game by Meg from the Teacher Studio). To play, you tell stories about animals on a farm. For each story, students kinesthetically act out what's happening. For example, say, "On the farm, there was 1 group of 3 chickens. How many chickens were there in all?" You would then have 3 students stand together as a group. This activity can be used for any sets of numbers, but since this section is on ×0 and ×1, you can continue to make up stories of various amounts involving zero groups of various sizes, various amounts of groups each having 0, 1 group of various sizes, and many groups each having 1. This activity may also be done using cubes or other small objects to model the various stories you tell. Along the way, ask students what they notice about the products. This will help them develop the conceptual understanding of the multiplication properties of 0 and 1.

It is important that students explore multiplication throughout the year in whole-group rich tasks, targeted small-group lessons, purposeful practice workstations with partners or independently, as well as at home with their guardians. These explorations should include continued practice with the concepts through games, energizers, routines, and problem solving (see Figures 3.7–3.15).

MATH WORKSTATIONS

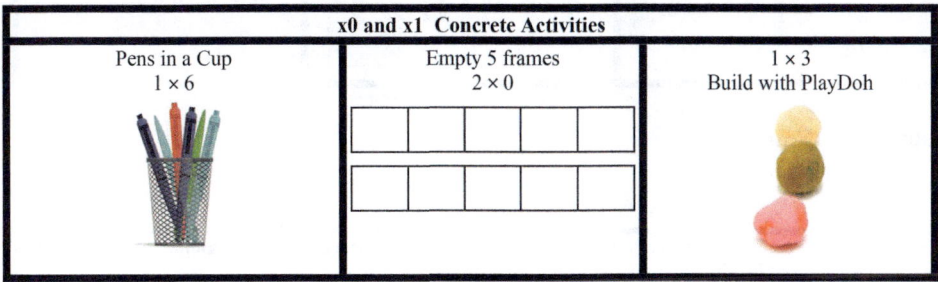

FIGURE 3.7 Concrete Activities

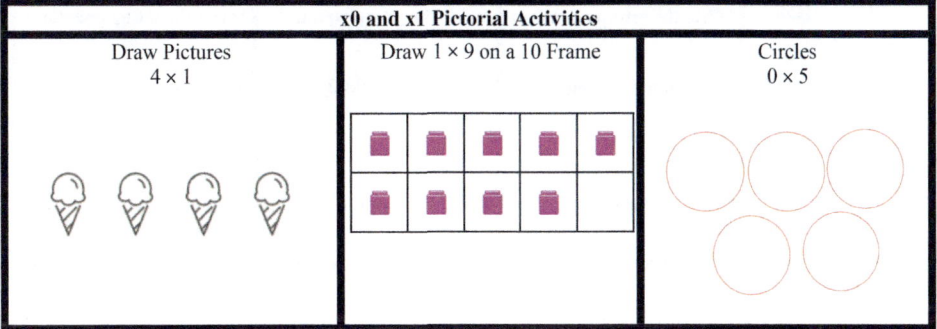

FIGURE 3.8 Pictorial Activities

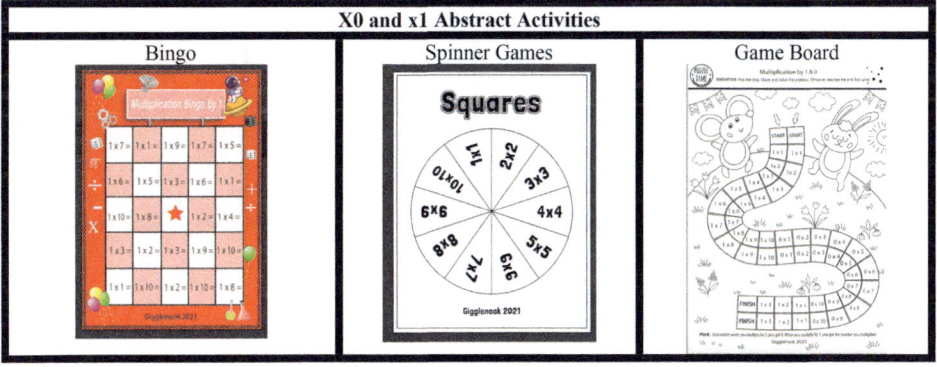

FIGURE 3.9 Abstract Activities

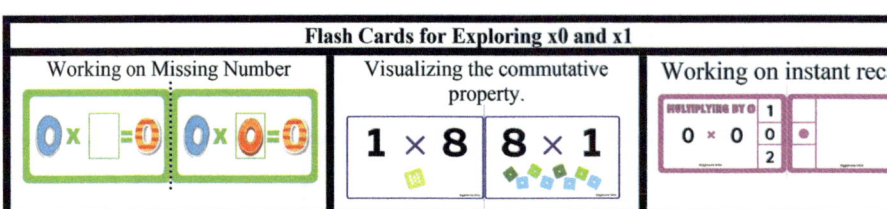

Flash Cards for Exploring x0 and x1		
Working on Missing Number	Visualizing the commutative property.	Working on instant recall.
$0 \times \square = 0$ $0 \times 0 = 0$	1×8 8×1	MULTIPLYING BY 0 0×0 0 1 0 2 •

FIGURE 3.10 Flash Cards

In every fluency module there should be a focus on word problems. Here are a few examples of the types of word problems that involve x0 and x1			
My x0 and x1 Story Problems Booklet	There were 3 goldfish bowls that were empty. How many goldfish were there? Write the set-up equation: Show your thinking with a model. Write the solution equation.	If one bag of marbles contains 8 marbles, how many marbles are there in all? Write the set-up equation: Show your thinking with a model. Write the solution equation.	A garden had one row of tomato plants. If there were 5 plants in each row, how many tomato plants were there? Write the set-up equation: Show your thinking with a model. Write the solution equation.

FIGURE 3.11 Story Problem Booklets

RESOURCES

Books about x0 and x1			
Stuart J. Murphy Too Many Kangaroo Things To Do https://www.youtube.com/watch?v=qq13df6XTo4	Neuschwander, Cindy Amanda Bean's Amazing Dream https://www.youtube.com/watch?v=FUvoeXpYwVQ	Angeline Lopresti A Place for Zero https://www.youtube.com/watch?v=41QBp1jTaYs	Joan Holub Zero the Hero https://www.youtube.com/watch?v=Kw2b6sDdPIk

FIGURE 3.12 Books

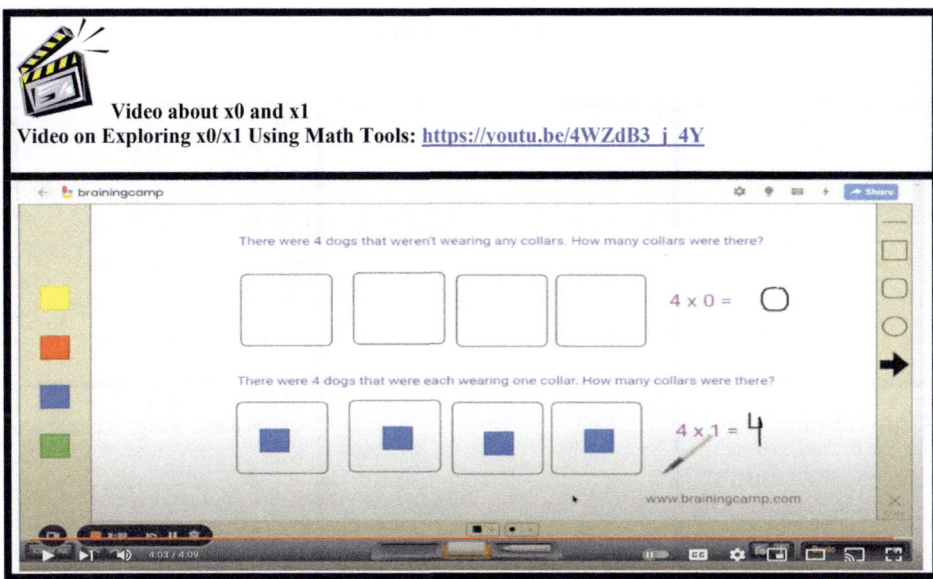

FIGURE 3.13   Videos

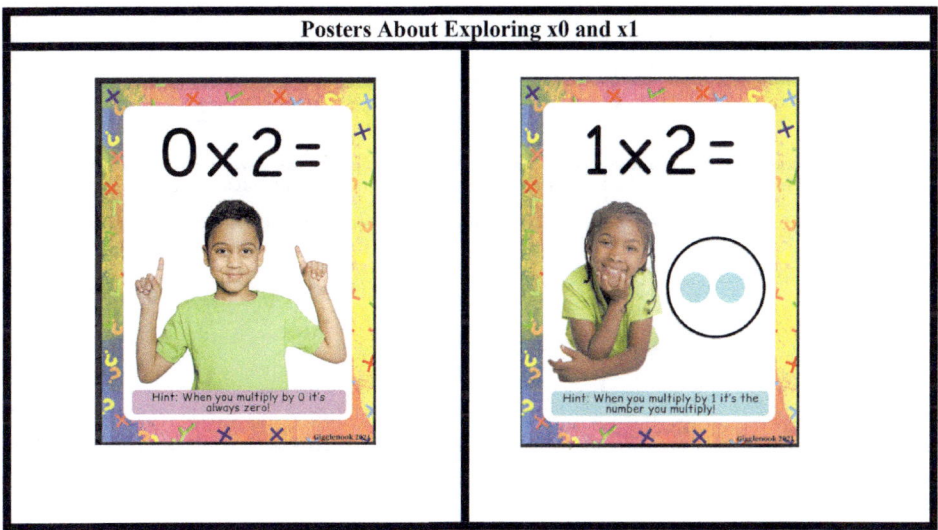

FIGURE 3.14   Posters

Multiplication by 0 and 1 Quiz

Name:

Date:

$0 \times 2 =$ Draw a picture.	$5 \times 1 =$ Draw as an array.	Solve. Kathryn bought her kitten 1 package of toys that had 4 toys in it. How many toys did her kitten receive? Write the equation: Strategy: _____ Answer: _____
Solve. Kylynn had 4 head bands that had 0 beads on them. How many beads were there? Write the equation: Strategy: _____ Answer:	Solve. $6 \times$ ____ $= 0$ ____ $\times 3 = 1$ $1 = 9 \times$ _____ ____ $= 0 \times 7$	What is the x0 strategy? Explain with numbers, words, or pictures

$5 \times 1 =$

Model on the number line.

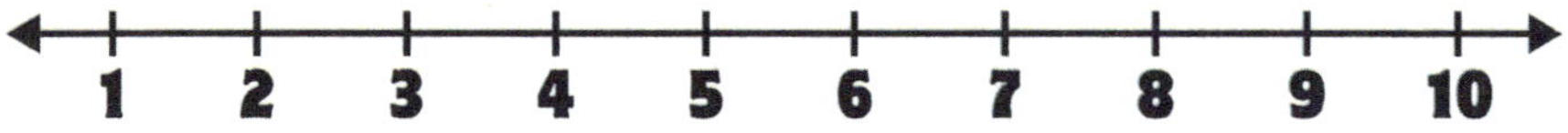

What happens when we multiply by a 1? Explain with numbers, words and pictures.

Circle how good you think you are at doing x0 and x1 facts!

Great	Good	Ok, still thinking

FIGURE 3.15 Quiz

EXPLORING AND LEARNING ×10 AND ×5 FACTS

One of the foundations of building number sense is using the benchmarks of 5 and 10 (Van De Walle et al., 2013) as we explore the relationship between 5 groups of objects and 10 groups of objects. Within the base 10 number system, there are patterns that emerge when we multiply by 10's. Students pick up on those patterns relatively easily and can solve 8 × 10 or 10 × 3 even if they don't truly understand why those patterns are happening. Perhaps our students have heard the seemingly magical recipe of "just adding a zero." If we stop and think about it, adding a zero makes no sense!!! If we were to "add a 0," the value of the number wouldn't change. What is actually happening is that the digits are getting 10 times greater, and this is reflected in place value position. If we take the time to explore this, then when students begin multiplying decimals by 10, they are set up for success.

It is crucial, then, that we have students *explain their thinking in mathematically precise ways* so that we can determine if they truly understand the patterns or are instead reciting rules that expire (Karp et al., 2014).

So often with the students we interview, they skip count by 5's to determine their ×5 facts. Even if they are multiplying a larger number by 5, they begin at 5. To facilitate efficient thinking, we can encourage students to use facts they know to determine the ones they don't yet know. For example, if they learn that 5 × 5 = 25, then they can start at 25 and then count by 5's from there. While this is still additive reasoning, it is the beginning of students thinking more efficiently and going beyond thinking of each group as one unit that needs to be counted separately.

One of the most powerful strategies we can use with multiplication is to scale up one factor by an amount and then to scale down the other factor by that same amount. One early exploration of that can be described as double and halving. It is not limited to double and halving, though, since we can triple and third and quadruple and fourth, and so on. We just need to be scaling up one factor and scaling down the other factor in the same way. By exploring that 5 groups of objects are the same as half of 10 groups of objects, students will be moving from additive reasoning into multiplicative reasoning. Since making our thinking visible is crucial for all learners, area models are a fabulous visual model to demonstrate why we are able to use this strategy (see Figures 3.16 and 3.17).

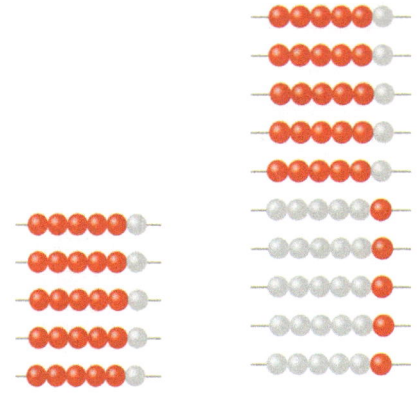

5 groups of 6   10 groups of 6

FIGURE 3.16 AND 3.17   Using the Rekenrek

It is worth taking time to explore this relationship with students since ×10 and ×5 products can be used so often to derive products and are not limited to basic facts. For example, 51 × 24. Students may want to rush to using an algorithm, but be aware of the thinking of the students as they do so. Too often, since the algorithms are robbed of place values, students are skip counting the basic facts within the calculations. Instead, what if we decompose the 51 into 50 groups of 24 and one more group of 24? The relationship of ×5 being half of ×10 can graduate to ×50 being half of ×100. So, if 100 × 24 = 2400 then I can reason that 50 × 24 = 1200. We only need one more group of 24 to go, so 51 × 24 must be 1224. We have arrived at the product using foundational facts. This is very different experience from the thinking involved in the standard algorithm, and so much more accessible to all of our students.

WHOLE-CLASS ACTIVITIES

Daily Routine—Number Strings

Number strings are a powerful daily routine that facilitates students using number relationships to determine products. The expressions in a number string are intended to help students discover that they don't need to skip count to determine their 5 facts but instead can use an easier fact of ×10 to determine the ×5 by halving it. The teacher will give each expression one at a time and facilitate the discussion of how knowing the ×10 can help determine the ×5. Here's a sample number string:

10 × 4
5 × 4
10 × 6

5 × 6
10 × 7
5 × 7

After recording the various products, ask students if they notice anything with the 10 facts and the 5 facts. They can discuss this with partners and then share out with the class.

WHOLE-CLASS MINI-LESSON

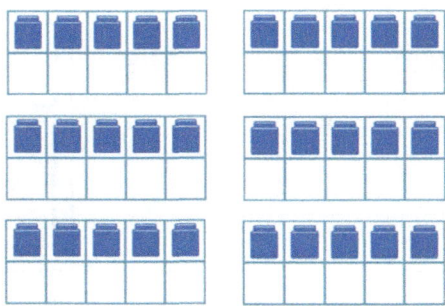

FIGURE 3.18

Launch: *Teacher projects the above image (see Figure 3.18) on the board and asks the students to determine how many cubes there are and how they determined how many there are.)*

Teacher: *I want everyone to have a chance to determine how many cubes are on the board, so put your thumb up in front of your chest when you can tell me how many and how you figured it out. For those of you who already figured it out, continue thinking about another way you could determine how many cubes there are. After waiting until all students have at least one thumb up, the teacher asks for volunteers.*

Bryan: *I think there are 30 cubes. I skip counted by 5's.*

Teacher: *Can you skip count out loud for us?*

Bryan: *5, 10, 15, 20, 25, 30*

Teacher: *Thank you, Bryan. Did anyone figure it out another way?*

Aiden: *I sort of did the same strategy as Bryan, but I didn't start with 5. I know that 4 × 5 = 20 so I then skip counted by 5's from the 20. 20, 25, 30, 35*

Teacher: *Thank you, Aiden. Did anyone think of it another way?*

Hank: *I skip counted going down the 10 frames, so 5, 10, and 15 and then saw that it was just the same amount again, so I added 15 + 15 and got 30.*

Teacher: *Thank you, Hank. Did anyone do it another way?*

Jackie: *I was thinking about the full 10 frames. If they were all full, then that would be 6 groups of 10, so 60. They are only half full, though, so I can figure out half of the 60. So, I know there are 30 cubes.*

Teacher: *There are so many different ways for us to see this amount and determine how many there are. I appreciate you all sharing your thinking with us.*

SPOTLIGHT ACTIVITY

Read the story *One is a Snail, Ten is a Crab* (April Putley Sayre and Jeff Sayre) and then make a variety of activities based on the book (see Figure 3.19).

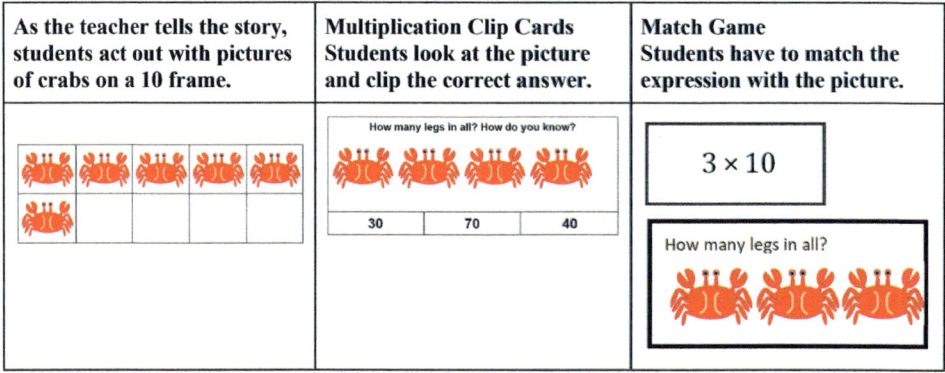

As the teacher tells the story, students act out with pictures of crabs on a 10 frame.	Multiplication Clip Cards Students look at the picture and clip the correct answer.	Match Game Students have to match the expression with the picture.

FIGURE 3.19 Spotlight Activity

The relationship between ×10 and ×5 is a powerful one not only for basic facts but throughout our students' math journeys. It is important that students get practice applying this relationship in a variety of activities and games. Here are some ideas for you as you provide opportunities both inside and outside your classroom (see Figures 3.20–3.29).

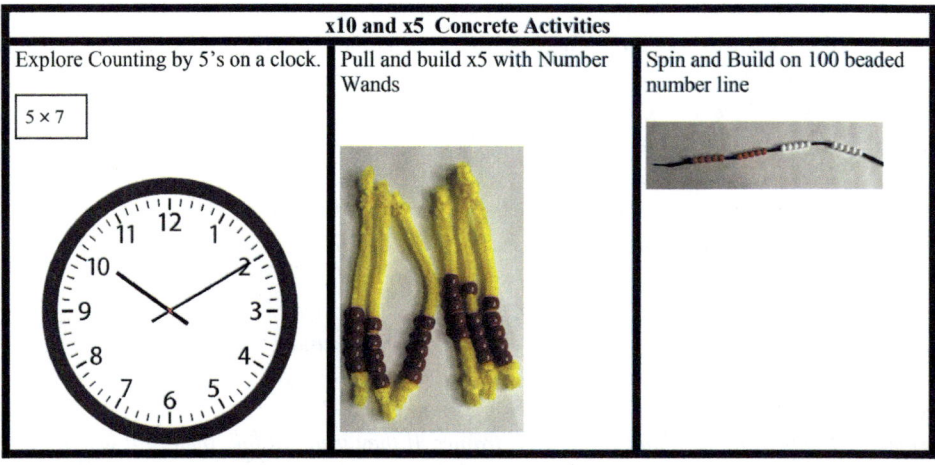

x10 and x5 Concrete Activities		
Explore Counting by 5's on a clock.	Pull and build x5 with Number Wands	Spin and Build on 100 beaded number line

FIGURE 3.20 Concrete Activities

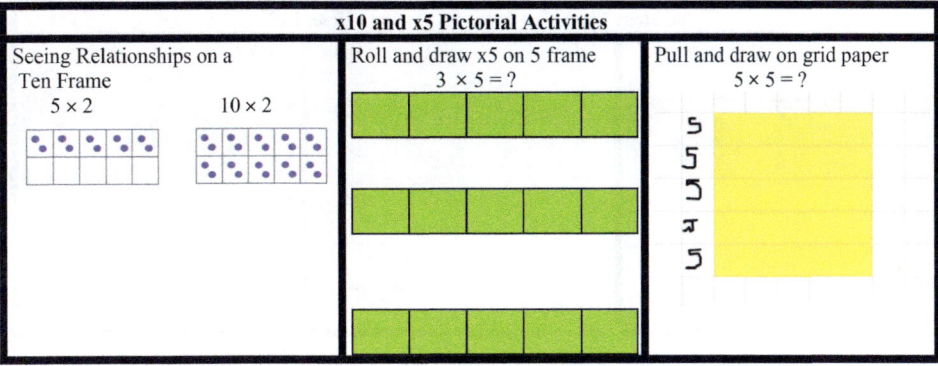

FIGURE 3.21 Pictorial Activities

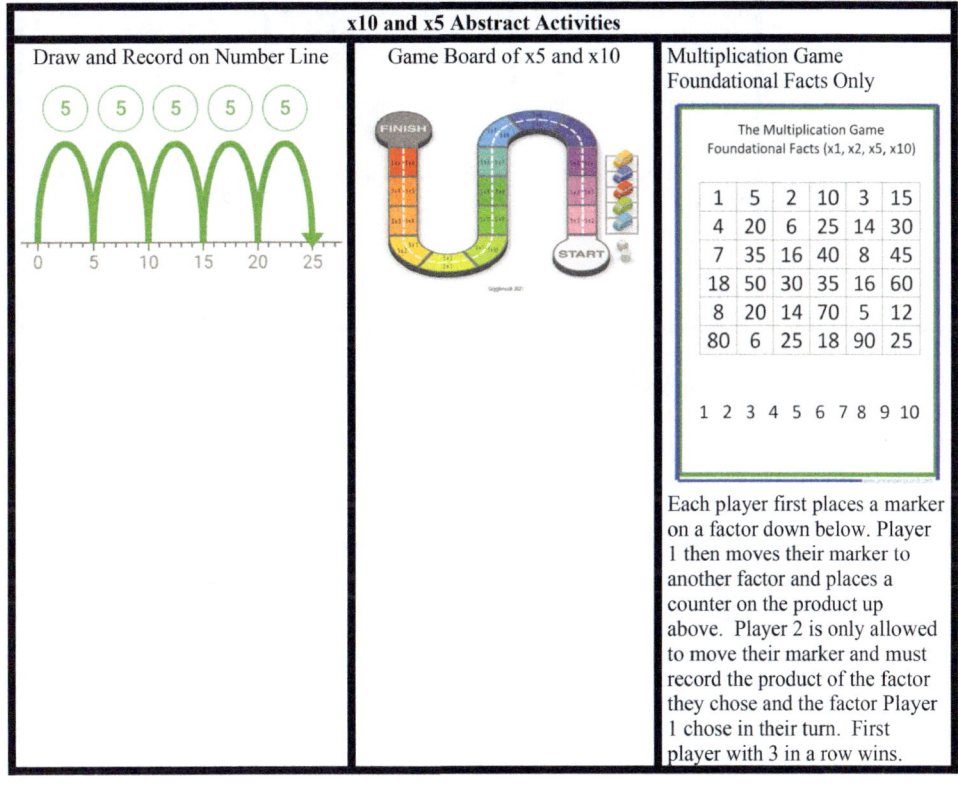

FIGURE 3.22 Abstract Activities

Strategy Flash Cards for Exploring x10 and x5		
Visualizing the Distributive Property	Visualizing Equal Groups and Arrays	Visualizing the Communitative Property

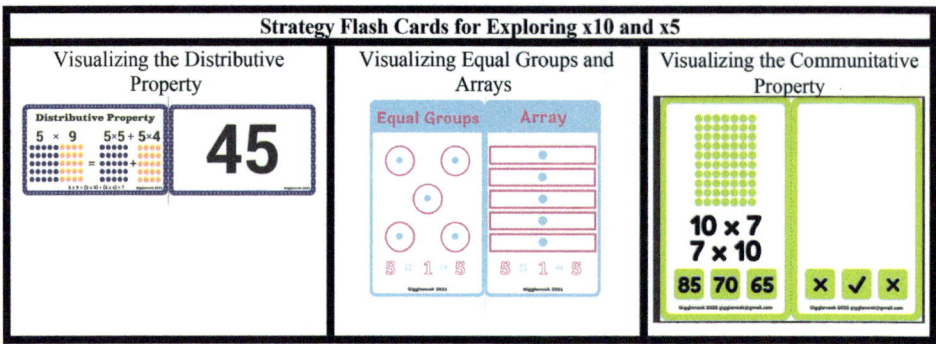

FIGURE 3.23 Strategy Flash Cards

In every fluency module there should be a focus on word problems. Here are a few examples of the types of word problems that involve x10 and x5			
My x10 and x5 Story Problems Booklet	If 6 trees each had 5 apples, how many apples were there in all? **Write the set-up equation:** **Show your thinking with a model.** **Write the solution equation.**	If each lane had 10 bowling pins, how many bowling pins are there in 8 lanes? **Write the set-up equation:** **Show your thinking with a model.** **Write the solution equation.**	Kyleigh ran 5 miles each day for one week. How many miles did she run? **Write the set-up equation:** **Show your thinking with a model.** **Write the solution equation.**

FIGURE 3.24 Story Problem Booklets

RESOURCES

Books about x10 and x5			
One is a Snail, Ten is a Crab by April Pulley Sayre and Jeff Sayre	Cindy Neuschwander Sir Cumference and All the King's Tens https://youtu.be/Fb2p2nMQbUw	Angeline Lopresti A Place for Zero https://www.youtube.com/watch?v=41QBp1jTaYs	Rey, H. A. Curious George: The Donut Delivery (younger story about power of 10 just to hundreds) https://youtu.be/nwHhrCDoZFo

FIGURE 3.25 Books

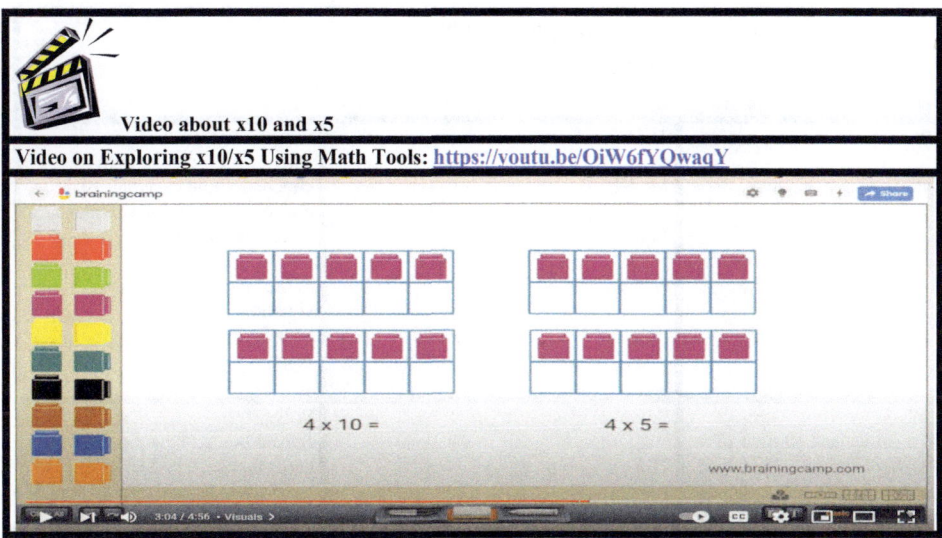

Video about x10 and x5

Video on Exploring x10/x5 Using Math Tools: https://youtu.be/OiW6fYQwaqY

FIGURE 3.26 Video

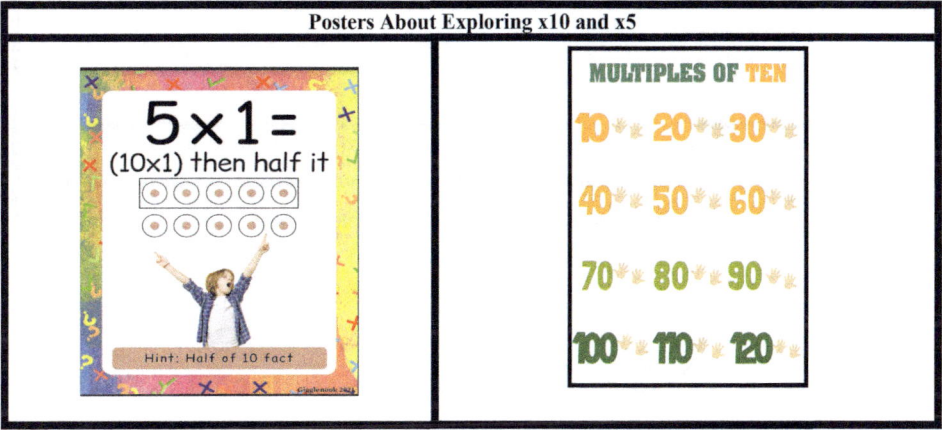

Posters About Exploring x10 and x5

FIGURE 3.27 Posters

Multiplication by 10 Quiz

Name
Date:

$10 \times 3 =$ Model with a drawing.	There are 5 cars. How many wheels? Model: Equation: Answer:	Solve. There were 5 packages of chocolates that each had 10 chocolates. How many chocolates were there in all? Write an equation for the situation: Strategy: _____ Answer: _____
Solve. If a garden had 10 rows of carrot plants with 7 plants in each row, how many carrot plants are there in all? Write an equation for the situation: Strategy: _____ Answer: _____	Solve. $10 \times$ ____ $=60$ ____ $\times 10 = 90$ $40 = 10 \times$ _____ _____ $= 10 \times 2$	Solve. Steven bought 8 packages of cards that each had 10 cards. He gave away 30 cards. How many cards did he have left? Write an equation for the situation: Strategy: _____ Answer: _____

$4 \times 10 =$
Model on a ten frame.

What pattern do you notice when you multiply by a 10? Explain with numbers, words and pictures.

Circle how good you think you are at doing x10 facts!

Great   Good    Ok, still thinking

FIGURE 3.28   Quiz on ×10

Multiplication by 5 Quiz Name Date:		
$5 \times 2 =$ Model with a drawing.	$5 + 5 + 5 + 5 + 5 + 5 + 5$ Write _____ × _____ = _____	Solve. A garden had 4 rows of bushes. If there were 5 bushes in each row, how many bushes were there? Write an equation for the situation: Strategy: _____ Answer: _____
Solve. An art show had 5 displays that each had 7 sculptures. How many sculptures were there in all? Write an equation for the situation: Strategy: _____ Answer: _____	Solve. $6 \times$ ____ $=60$ ____ $\times 5 = 30$ $50 = 10 \times$ _____ _____ $= 5 \times 8$	Solve. The first day Manny did sit-ups, he could do 8 sit-ups without stopping. After practicing every day, one month later he was able to do 5 times as many sit-ups. How many situps was he now able to do? Write an equation for the situation: Strategy: _____ Answer: _____
$4 \times 10 =$ Write a word problem. Model and solve.		
What happens when we multiply by a 10? Explain with numbers, words and pictures.		
Circle how good you think you are at doing x5 facts! Great Good Ok, still thinking		

FIGURE 3.29   Quiz on ×5

EXPLORING AND LEARNING ×2, ×4, AND ×8 FACTS

The next set of facts we can explore includes the related doubles facts. We have interviewed many students who struggled with addition within 10 facts but knew all their doubles facts. By exploring the ×2 facts at this point, we can more easily connect those facts to the ×4 and ×8 facts by using visual representations such as 10-rowed rekenreks, Cuisenaire® rods, or arrays. There are a variety of ways that students may decompose 4 groups and 8 groups, but for those who are still in additive reasoning and skip counting, relating these facts to thoughts of doubling is a wonderful entry into the use of partial products that are already known. A word of caution, though: we always want to be mindful that we aren't telling students "when you see ×4, then use double." Instead, we need students to develop that understanding on their own. Our ultimate goal is that students develop flexibility with decomposing factors in a variety of ways or use number relationships to determine products. If we set this foundation of understanding, students won't just master their ×2, ×4, and ×8 facts; they will be able to multiply anything times 2, 4, or 8.

Many upper elementary students continue to repeatedly add numbers when asked a multiplication problem, even when there are larger numbers. Sometimes, they will group two numbers together until they have actually halved the number of groups and doubled the size in each group—a really powerful multiplication strategy that can apply to all sets of numbers. When this happens, it can be an ideal time to introduce the doubling and halving strategy and make it visible to all students. Here is a link to a sample: https:// youtu.be/TOZIsiYfGYY.

For those who wonder how we can explicitly help students move from additive thinking to multiplicative thinking, knowing this teaching move can make a huge difference. If you happen to see this with your own students, be sure to wonder aloud, "So you are saying that 8×8 is the same 4×16, which is the same as 2×32?" even if they haven't written any multiplication problems on their work. Then, have them look at those three expressions and ask them what they notice. It's so much more impactful when the relationships are recognized by the students rather than us telling them.

WHOLE-CLASS ACTIVITIES

Routines—Double and Halving

As a daily routine, you can call out a number and have the students determine either the double or the half of that number, always explaining their strategies for doing so. In this way, they will build these powerful relationships.

WHOLE-CLASS MINI LESSON

Introduction: *(Teacher will show a classroom-sized 10-rowed rekenrek that has 8 rows of 6 beads slid over to the left (see Figure 3.30).)*

FIGURE 3.30   Rekenrek 8 × 6

Teacher: *I would like you all to determine how many beads are on the left side of this reken-*
rek. I will give you all a minute of private think time. When you know how many
there are and can explain the strategy you used, put a thumb up in front of your
chest, and while you are waiting, think of another strategy you could use.
Teacher will then provide think time so that all students have a thumb up in front
of their chest.

Vibianna: *I think there are 48 beads. I skip counted by 5's until I got to 40, and then I*
thought 41, 42, 43, 44, 45, 46, 47, 48.

Teacher: *I'm going to move these beads so that Vibianna's thinking is visible to us all.*
Teacher will move the extra bead away from the groups of 5 to show how Vibi-
anna broke apart the groups into 5 groups and 1 more group (see Figure 3.31).

FIGURE 3.31   Rekenrek Showing (8 × 5) + (8 × 1)

Teacher: *Thank you, Vibianna. Did anyone use a different strategy?*

Justin: *I broke it apart the same way that Vibianna did, but I knew that 8 × 5 = 40 and*
then I added 8 for the 48.

Teacher: Thank you, Justin. It's so helpful when we can use facts that we know to help us determine the ones we don't yet know. Did anyone use a different strategy?

Kobe: I know that 4 groups of 6 is 24, so I just saw that the amount that is left is another 4 groups of 6. So, I doubled the 24 to 48.

Teacher: Let me move these beads so I can show your thinking, Kobe. Let me know if I'm not modeling it the way you thought about it (see Figure 3.32).

Teacher demonstrates this on the rekenrek.

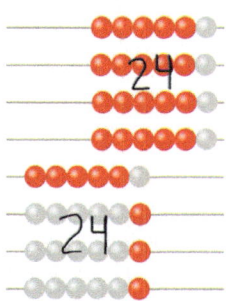

FIGURE 3.32 Rekenrek Demonstrating ×4 to ×8

Teacher: Thank you, Kobe. Did anyone use another strategy?

Martha: I knew that 2 × 6 = 12 so I doubled that to get 4 × 6 = 24. Then I doubled the 24 to 48. So, I used the double, double, double strategy.

Teacher: So, I can move the beads to show your thinking, Martha. Is this how you thought of it? (See Figure 3.33.)

Martha: Yes, that's it.

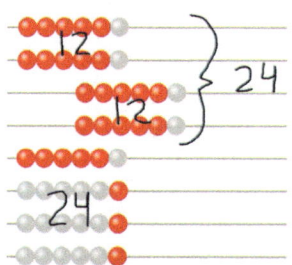

FIGURE 3.33 Rekenrek Demonstrating Double Double Double

Teacher: Thank you, everyone. There are so many different strategies we can use to figure out 8 × 6. I appreciate you all sharing your thinking with us!

SPOTLIGHT ACTIVITY

A wonderful way to begin exploring the idea of doubling is with the book *Two of Everything* by Lily Toy Hong. In the story, the Haktaks discover a magical pot that will double anything you put into it, including a hairpin, a purse with gold coins, a coat, and even themselves! As you read the story together, students can act out putting items in a pot and predicting how many items will come out after being doubled. You can also extend this by reversing the scenario. Share the amount that came OUT of the pot and have students determine how much must have gone in.

Another engaging book involving doubling is *Minnie's Pies: A Multiplication Menu* by Dale Ann Dodds (https://youtu.be/-t4mmOacKqY). Students can then create their own menu of items and explore what would happen if Millie doubled those items. This provides some differentiation for students to use numbers that are comfortable for them.

Some students are automatic with facts up to the ×4's. At this point, they typically skip count by one of the factors. Others begin to use the doubling thinking, but their additive doubling skills are not strong yet. In this case, students need purposeful practice to choose to use a strategy and develop efficiency with using it. Here are lots of ideas to help students explore these relationships (see Figures 3.34–3.44).

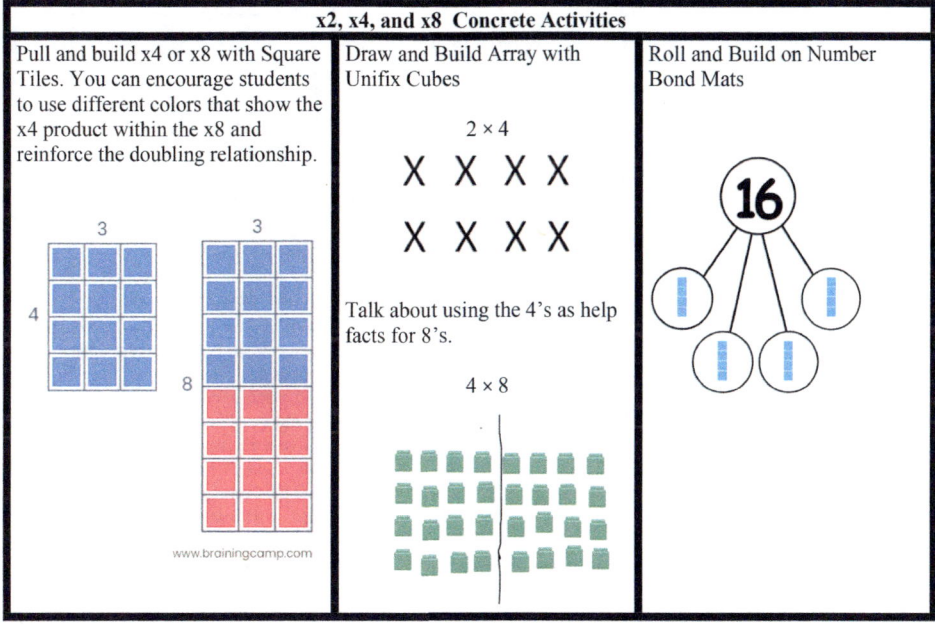

FIGURE 3.34 Concrete Activities

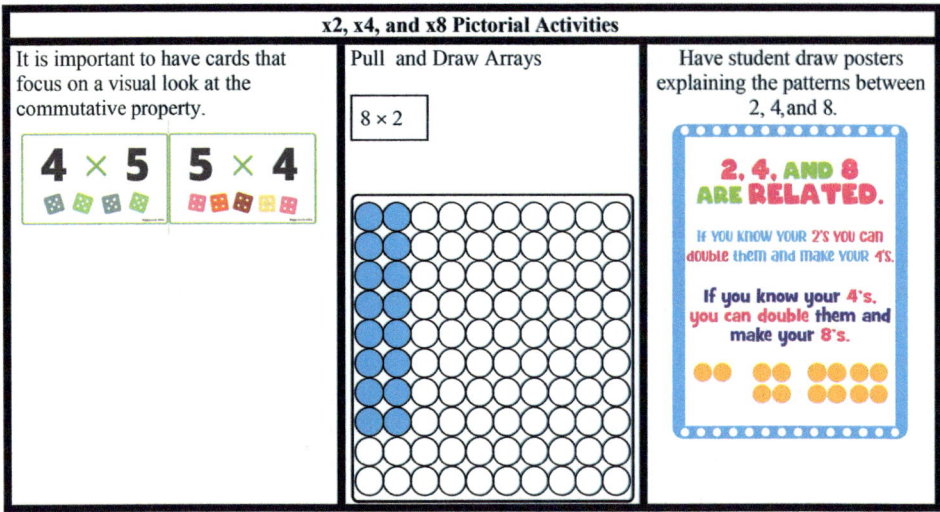

FIGURE 3.35 Pictorial Activities

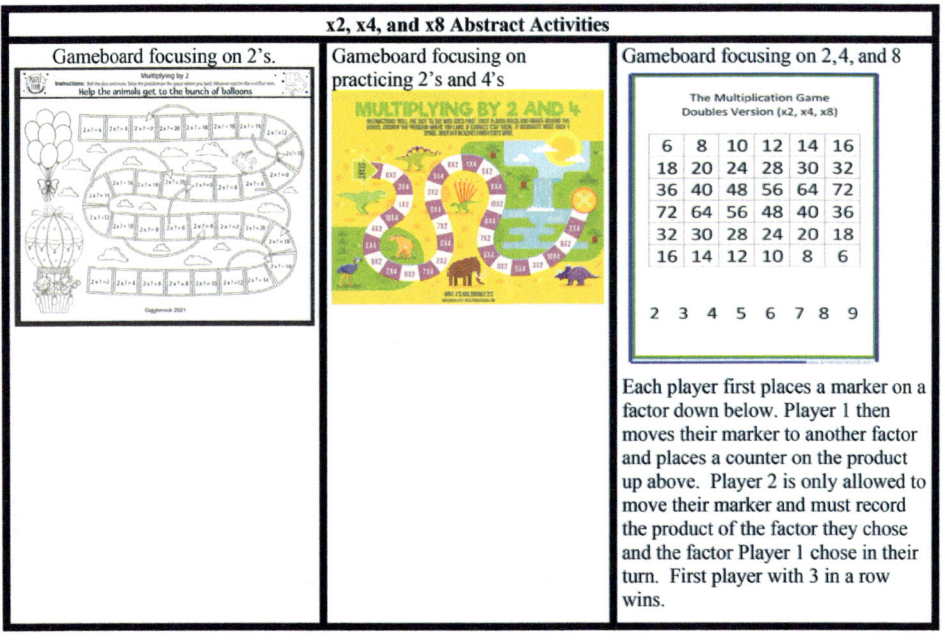

FIGURE 3.36 Abstract Activities

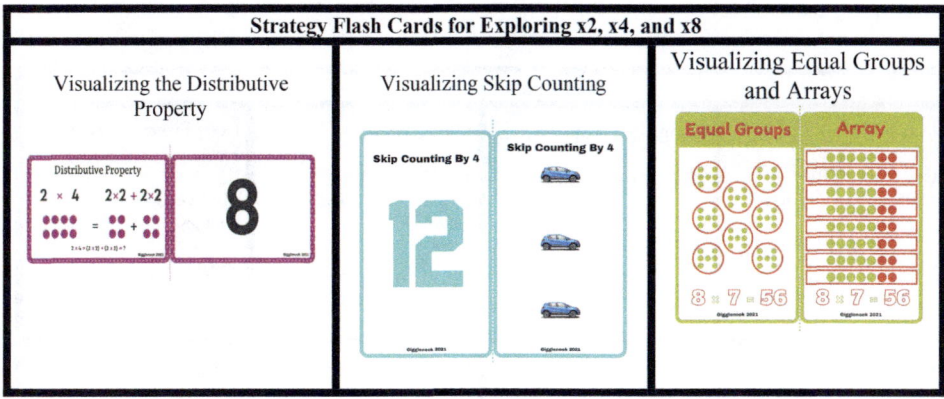

FIGURE 3.37 Strategy Flash Cards

	In every fluency module there should be a focus on word problems. Here are a few examples of the types of word problems that involve x2, x4, and x8		
My x2, x4 and x8 Story Problems Booklet	Matthew owned 6 board games. Hannah owned 4 times as many as Matthew. How many did Hannah own?	Kathryn had 4 cups of paint brushes. If each cup had 7 paint brushes, how many paint brushes did she have?	At a pet store, there were 3 3 rows of fish tanks. If each row had 8 fish tanks, how many fish tanks were there?
	Write the set-up equation:	**Write the set-up equation:**	**Write the set-up equation:**
	Show your thinking with a model.	**Show your thinking with a model.**	**Show your thinking with a model.**
	Write the solution equation.	**Write the solution equation.**	**Write the solution equation.**

FIGURE 3.38 Story Problem Booklet

RESOURCES

Books about x2, x4, and x8			
Demi One Grain of Rice: A Mathematical Journey https://youtu.be/K6GuLwSJTIY	Hong, Lily Toy Two of Everything https://www.youtube.com/watch?v=JML_7tsqlmU	Elinor Pinczes 100 Hungry Ants (discuss double and halving) https://www.youtube.com/watch?v=8qhUaQpaRrQ	Doris Fisher and Dani Sneed My Even Day (lower doubles) https://www.youtube.com/watch?v=Bk4HftjD8PQ
Kathryn Cristaldi Even Steven Odd Todd (even/odd) https://www.youtube.com/watch?v=AA02JTIN6gA	Suzanne Aker What Comes in 2's, 3's and 4's? https://www.youtube.com/watch?v=c6_lhuYDoOs	Dodds, Dayle Ann Minnie's Diner (doubling) https://www.youtube.com/watch?v=-t4mmOacKqY	Ghigna, Charles One Hundred Shoes: A Math Reader (double and halving) https://www.youtube.com/watch?v=ax4akgElTGA
Brenner, Martha F. Stacks of Trouble: Multiplication https://www.youtube.com/watch?v=uYdMXATScNM	Shaskan, Trisha Speed If You Were a Times Sign https://www.youtube.com/watch?v=ckmMbQ5DM5M		

FIGURE 3.39 Books

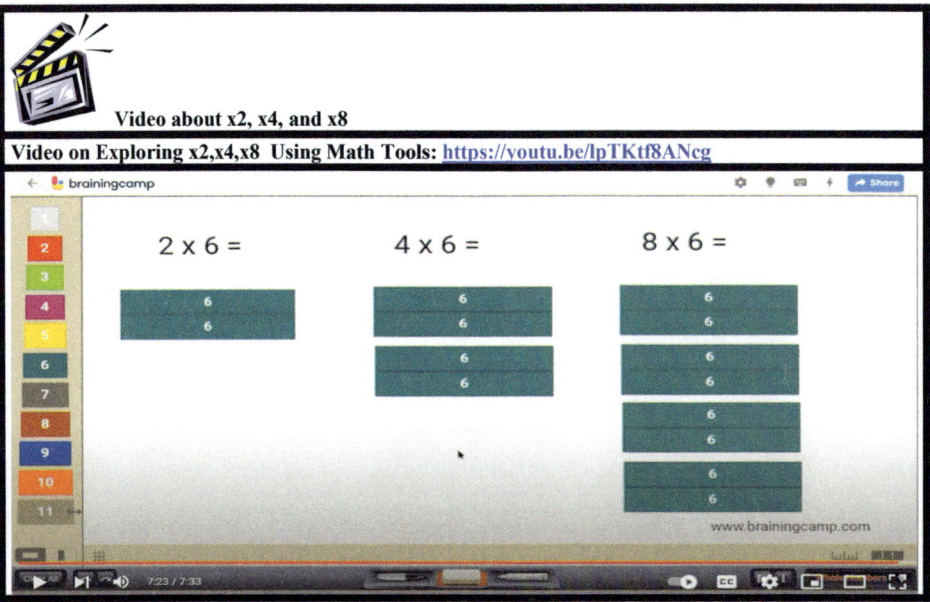

FIGURE 3.40 Video

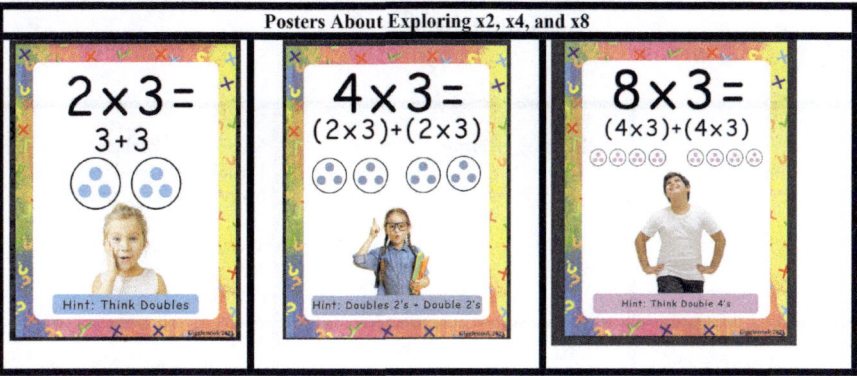

FIGURE 3.41 Posters

Multiplying by 2 Quiz		
Name		
Date:		
$2 \times 4 =$ Model using equal groups.	$2 \times 5 =$ Model using an array.	Solve. A recipe will make 8 muffins. How many muffins can be made with 2 recipes? Write the equation: Strategy: _____ Answer: _____
Solve. Vibiana owns 9 pairs of shoes. How many shoes does she own in all? Write the equation: Strategy: _____ Answer: _____	Solve. $2 \times$ ____ $= 12$ $7 \times 2 =$ _____ $18 = 2 \times$ _____ _____ $= 2 \times 6$	Solve. A fruit bowl had some apples and oranges. If there were 3 oranges and 2 times as many apples than oranges, how many apples were there in the bowl? Write the equation: Strategy: _____ Answer: _____

$2 \times 8 =$

Model on a ten frame.

What is a strategy we can use when we multiply by an 2? Explain with numbers, words and pictures.

Circle how good you think you are at doing x2 facts!
  Great   Good   Ok, still thinking

FIGURE 3.42 Quiz ×2

Multiplying by 4 Quiz

Name
Date:

$7 \times 4 =$ Model showing equal groups.	$4 \times 8 =$ Model using an array.	Solve. Marty had 4 baskets of dog toys. If each basket had 9 toys, how many toys did he have? Write the equation: Strategy: _____ Answer: _____
Solve. Jen had a bookshelf that had 6 shelves. If there were 4 books on a shelf, how many books were there in all? Write the equation: Strategy: _____ Answer:	Solve. $4 \times$ ____ $= 16$ $9 \times 4 =$ _____ $32 = 4 \times$ _____ _____ $= 7 \times 4$	Solve. The bakery made 8 batches of 10 cookies. They sold 50 cookies. How many cookies did they have left? Write the equation: Strategy: _____ Answer: _____

$4 \times 8 =$
Model in two different ways using the rectangles below.

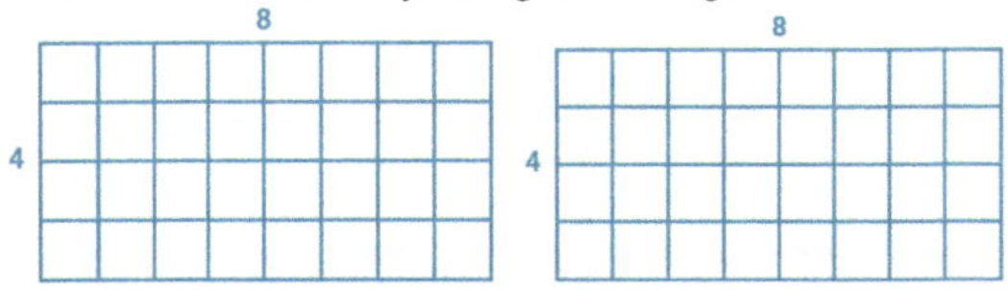

What is one strategy we can use when we multiply by an 4? Explain with numbers, words and pictures.

Circle how good you think you are at doing x4 facts!
Great   Good   Ok, still thinking

FIGURE 3.43   Quiz ×4

Multiplying by 8 Quiz

Name
Date:

$3 \times 8 =$ Model using equal groups.	$5 \times 8 =$ Model using an array.	Solve. One box of markers had 8 markers. How many markers are in 6 boxes? Write the equation: Strategy: _____ Answer: _____
Solve. The bakery had 8 boxes with 4 cookies in each box. They sold 5 boxes. How many cookies did they have left?d Write the equation: Strategy: _____ Answer: _____	Solve. $8 \times$ ____ $= 16$ $0 \times 8 =$ _____ $56 = 8 \times$ _____ _____ $= 8 \times 4$	Solve. Denise wanted to make 8 bracelets for her friends. If each bracelet required 8 beads, how many beads did she need? Write the equation: Strategy: _____ Answer: _____

$7 \times 8 =$

Model in two different ways using the rectangles below.

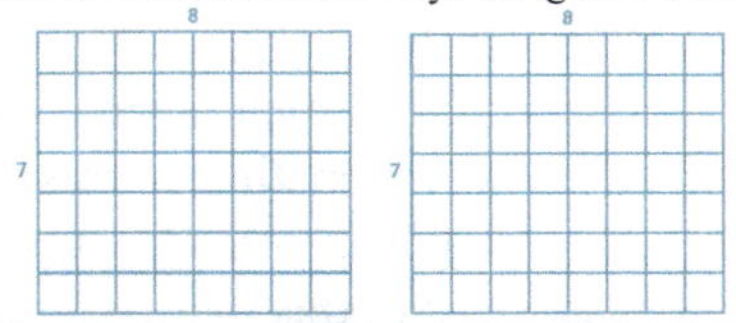

What is one strategy we can use when we multiply by an 8? Explain with numbers, words and pictures.

Circle how good you think you are at doing x8 facts!

Great   Good   Ok, still thinking

FIGURE 3.44 Quiz ×8

EXPLORING AND LEARNING ×3 AND ×6 FACTS

The ×3 facts appear at this point in the progression because we can explore the relationship of the ×6 facts being double the ×3 facts. If students have not mastered the ×3 facts yet, we can relate 3 groups to being 1 group more than the doubles of that factor. Once those are mastered, we can use math tools such as Cuisenaire® rods and 10-row reken-reks, or pictorial representations of quantities such as dot patterns on 10-frames or arrays, to connect the ×3 facts to the ×6 facts.

Another way of exploring ×6 facts is to break apart the 6 into 5 groups and 1 group. Once again, we are facilitating students' thinking into multiplicative reasoning and beyond skip counting by using foundational facts to derive higher facts. If we develop this flexibility, we can extend the idea to include place value while working with multi-digit numbers, empowering students to tackle any problems.

WHOLE-CLASS ACTIVITIES

Routines—Same but Different

Susan Looney has created a wonderful website (www.samebutdifferentmath.com) with images that encourage discussion about similarities and differences in math concepts for K–12. Here is an image that is perfect for discussing the relationship between ×3 facts and ×6 facts (see Figures 3.45 and 3.46) The website describes six steps to this routine.

1. Students look carefully at the image they are given.
2. Students have private think time to notice the similarities and differences between the photos.
3. Students turn and talk with a partner.
4. Students then share out to the whole class.
5. The teacher to helps consolidate the noticings of the students.
6. Students create an image of their own that shows the same concept conveyed in the pictures.

FIGURES 3.45 Plant Array 6 × 6

FIGURE 3.46   Plant Array (3 × 3) + (3 × 3)

(Sue Looney, 2022)

WHOLE-CLASS MINI LESSON

Teacher: (Inspired by Berkeley Everett's Math Flips) Today we are going to explore more relationships when we are multiplying. I'm going to show you an image, and I want you to think to yourself for a minute while you figure out how many dots there are.

Teacher shows the following image:

A

FIGURE 3.47   Dice: 3 Groups of 5

Deb: *I knew there were 15 dots because I know that 3 × 5 = 15.*
Teacher: *That's wonderful that you know that. What if a friend of yours didn't know that. How would you help them figure it out?*
Deb: *I would tell them to skip count by the number that isn't 3.*
   Teacher models that thinking on the board on a number line.
Teacher: *Thank you, Deb. Does anyone have another strategy when they multiply by 3?*
Jeff: *When I'm multiplying by 3, if I don't know it right away, then I think of the doubles fact first and then I add on one more group of the number being multiplied by 3.*
   Teacher models Jeff's thinking using equations.
Teacher: *Even though we all thought of different strategies for 3 × 5, we did all come to the conclusion that 3 × 5 = 15. I'm going to show you another image, and I want you think about how knowing 3 × 5 = 15 can help you figure out how many dots are in this next image.*

B

FIGURE 3.48   Dice: 6 Groups of 5

(Teacher provides personal think time and then asks students to turn and talk to a partner to share any strategies.)

Ben: *My partner and I think that there are 30 dots because if you look at the first image of the 3 dice, those are in this new image, but it is there twice.*

Teacher: *Does anyone else see it like that? Can you visualize that these 6 groups of 5 is double of the 3 groups of 5? It looks like it is double the first image. What might that mean about the ×3 and ×6 facts?*

Deb: *For ×6 facts, it seems way easier to double the ×3 fact than it is to skip count all the way up to the answer, so if I had to help a friend to figure out 6 × 5, I would ask them if the know what 3 × 5 is and if they do, they could just double it.*

Teacher: *Yes, I certainly agree, Deb. I like how you didn't give them an answer but instead were suggesting a way they could use what they know to figure it out for themselves.*

 SPOTLIGHT ACTIVITY

Image Talks

We absolutely love doing image talks with students because they offer access to students who may be intimidated by numbers. They simply see a picture and are asked to talk about what they notice, what they wonder, and what math might be involved. In this case, we are looking at multiplying by 6. Looking at eggs, something students have seen before, feels familiar and accessible. We show many different pictures (see a few herein) and ask students to name the equations, write the equations, and tell stories. Before you know it, we have found a way into thinking about and multiplying by 6 (see Figure 3.49)!

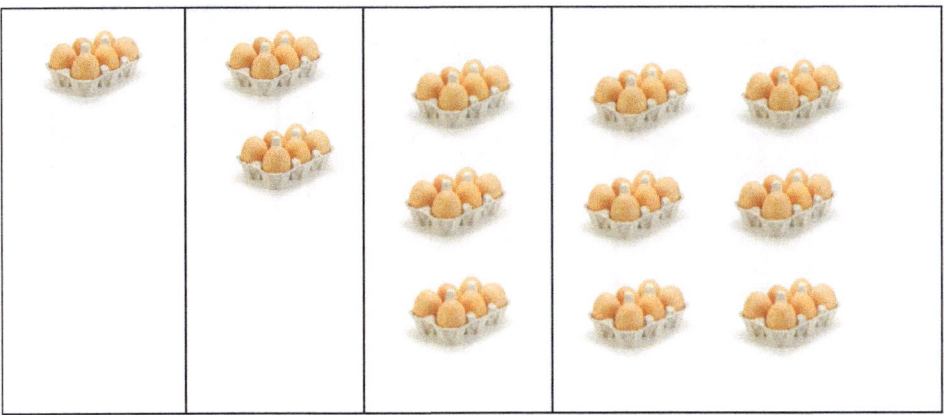

FIGURE 3.49 Image Talk

In the next figure, we illustrate this with a ratio table (inspired by Pam Harris www. mathisfigureoutable.com) (see Figure 3.50).

1 carton	2 cartons	3 cartons	6 cartons
6 eggs	**?**	**??**	**???**
	We ask how does what we know about 1 carton help us think about 2?	So then we say what could we do to figure out how many are in 3 cartoons.	What in this table could help us figure this out.

FIGURE 3.50 Ratio Table

Remember, we never want to dictate to students that they must use one strategy over another but instead provide them the opportunity to use multiplication strategies that work best for them. This is particularly true for ×6's. As we have discussed at the beginning of this chapter, we can relate the ×6 facts to the already-known ×3 facts or 6 groups as 5 groups plus 1 more group. Neither strategy is better than the other, but we do want students to be empowered to use the one that is most efficient for their brains. The following figures offer ideas to help students explore these relationships (see Figures 3.51–3.60).

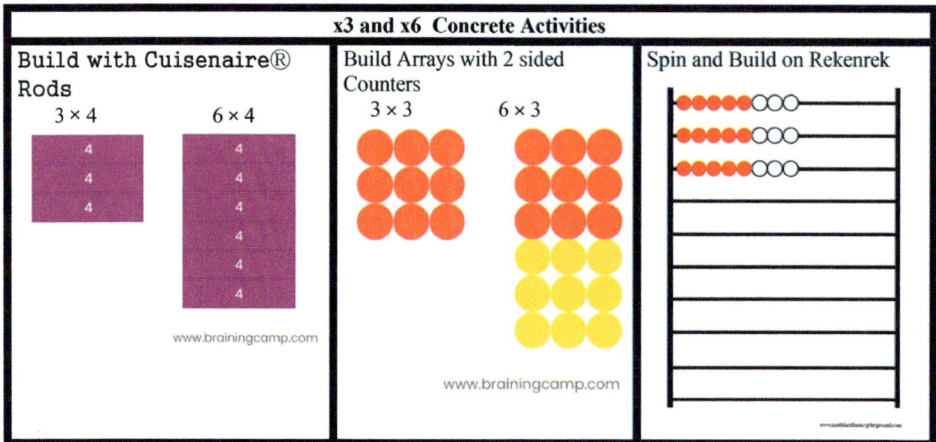

FIGURE 3.51 Concrete Activities

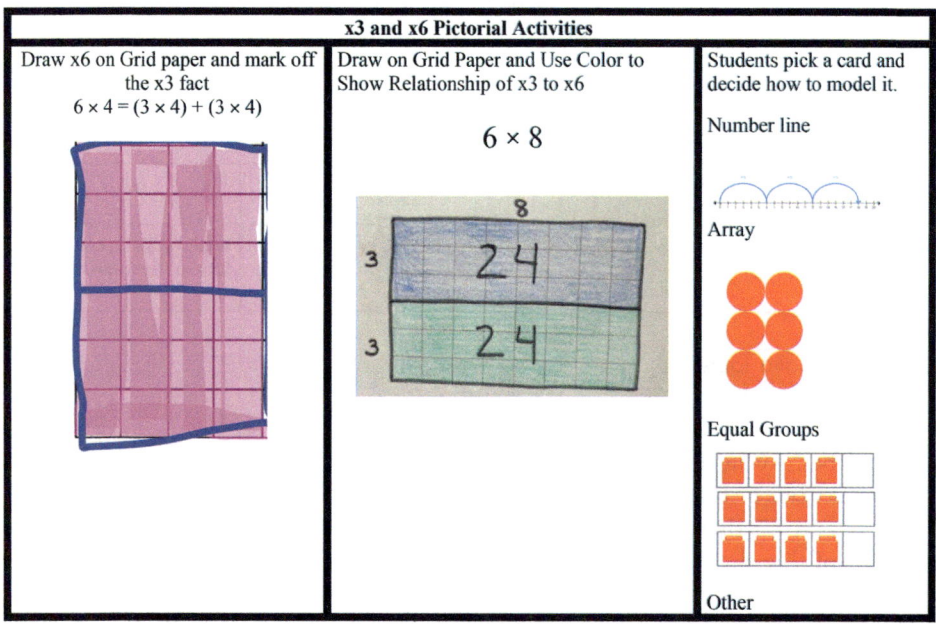

FIGURE 3.52 Pictorial Activities

x3 and x6 Abstract Activities		
Power Towers Write expressions on bottom of the outside and write the product on the inside for self-correcting. Differentiate by putting x3 and x6 facts with some of the previously mastered sets of numbers. 	4 in a Row 	Gameboard with both facts

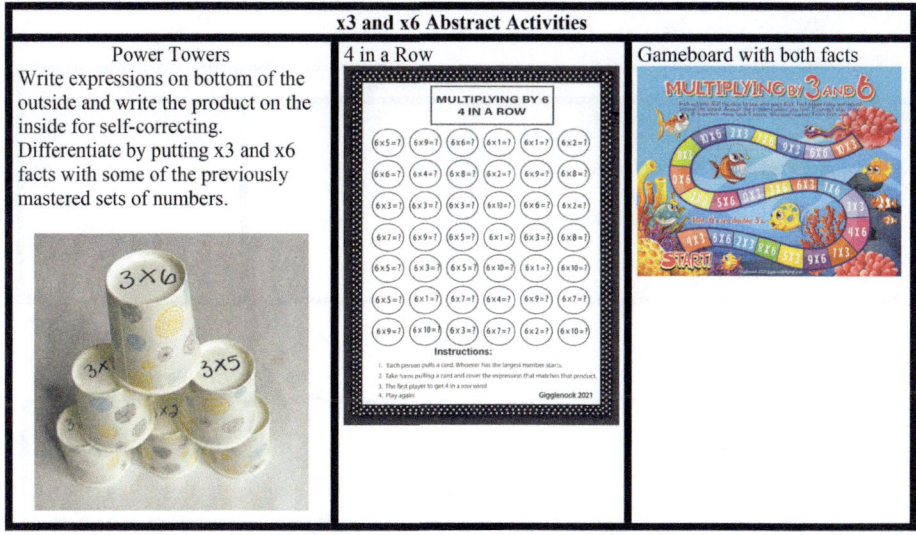

FIGURE 3.53   Abstract Activities

Strategy Flash Cards for Exploring x3 and x6		
Visualizing the Commutative Property	Visualizing the Commutative Property	Reasoning about Operations…Students have cards that are missing different operations and they have decide which operation makes sense

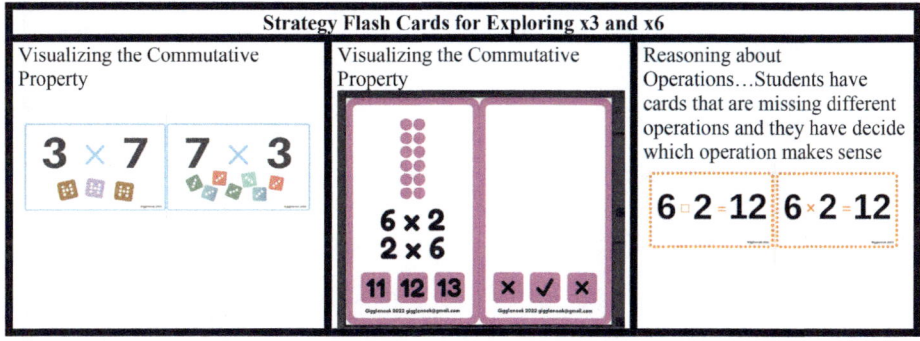

FIGURE 3.54   Strategy Flash Cards

	In every fluency module there should be a focus on word problems. Here are a few examples of the types of word problems that involve x3 and x6		
My x3 and x6 Story Problems Booklet	There were 3 trees with 7 birds in each tree. How many birds were there?	Sue collected 6 bags of sea shells. If each bag could fit 8 sea shells, how many sea shells does she have?	Jessica baked cookies and placed them on a tray. If there were 3 rows of cookies each having 6 cookies, how many cookies did she bake?
	Write the set-up equation:	Write the set-up equation:	Write the set-up equation:
	Show your thinking with a model.	Show your thinking with a model.	Show your thinking with a model.
	Write the solution equation.	Write the solution equation.	Write the solution equation.

FIGURE 3.55   Story Problem Booklet

RESOURCES

Books about x3 and x6			
Giganti, Paul Jr. Each Orange Had 8 Slices https://www.youtube.com/watch?v=2fYKMrCbaE8	Palotta, Jerry The Hershey's Milk Chocolate Multiplication Book https://www.youtube.com/watch?v=FfDIHTjOyz0		

FIGURE 3.56 Books

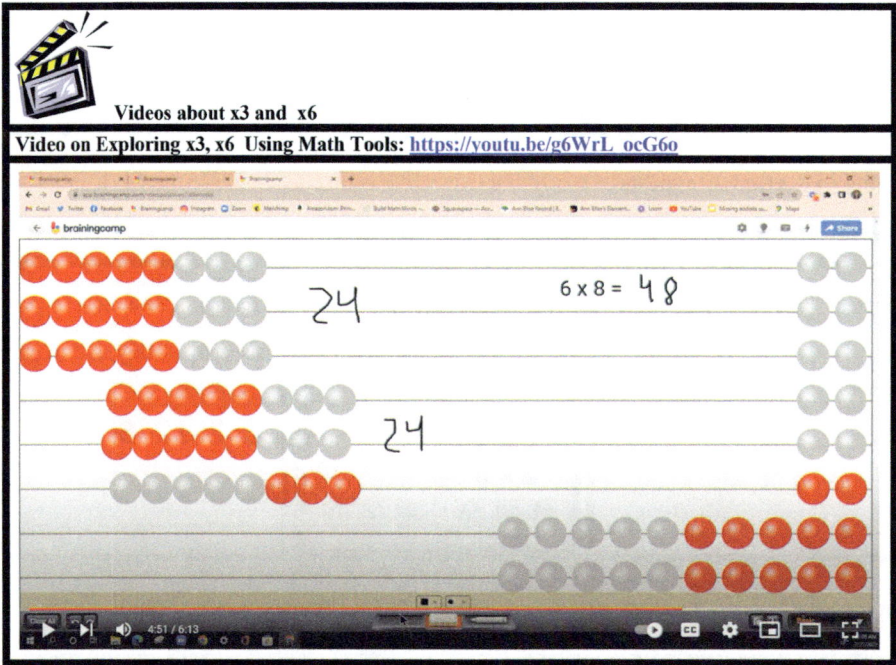

FIGURE 3.57 Video

FIGURE 3.58 Posters

Multiply by 3 Quiz

Name
Date:

$3 \times 7 =$ Model with equal group.	$5 \times 3 =$ Model with an array.	Solve. There were 4 tigers at the zoo. There were 3 times as many monkeys as tigers. How many monkeys were there? Write the equation: Strategy: _____ Answer: _____
Solve. If a bookshelf had 3 shelves with 8 books on each shelf, how many books were on the bookshelf? Write the equation: Strategy: _____ Answer:	Solve. $3 \times$ ____ $= 27$ ____ $\times 3 = 9$ $12 = 3 \times$ _____ ____ $= 3 \times 5$	Solve. There were 3 packages of chocolates which each had 9 chocolates. How many chocolates were there? Write the equation: Strategy: _____ Answer: _____

$3 \times 6 =$

Model using a tape diagram..

What strategy would you use for x3 facts if you didn't know one of them? Explain with numbers, words and pictures.

Circle how good you think you are at doing x3!

Great   Good   Ok, still thinking

FIGURE 3.59   Quiz ×3

Multiplying by 6 Quiz

Name
Date:

$6 \times 3 =$ Model with equal groups.	$4 \times 6 =$ Model on the area model.	Solve. If 6 puppies were each given 2 snacks, how many snacks were given to the puppies? Write the equation: Answer: _____
Solve. Christine's stamp collection was put in an album that fit 6 rows with 6 stamps in each row on one page. How many stamps can fit on each page? Write the equation: Answer: _____	Solve. $6 \times$ _____ $= 0$ _____ $\times 3 = 0$ $1 = 9 \times$ _____ _____ $= 1 \times 7$	Solve. One small package of gum balls had 9 gum balls. A large package had 6 times as many gum balls. How many gum balls were in the large package? Write the equation: Answer: _____

$6 \times 8 =$

Use two different colors to model 6 x 8 in two different ways using the area models below.

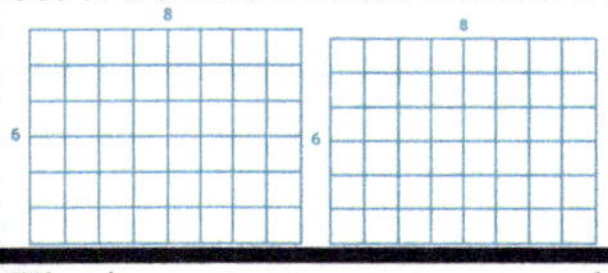

What is one strategy you can use when you multiply by a 6? Explain with numbers, words and pictures.

Circle how good you think you are at doing x6 facts!

Great   Good   Ok, still thinking

FIGURE 3.60 Quiz ×6

EXPLORING AND LEARNING ×9 FACTS

The ×9 facts are next in the progression and may be derived by combining the products of the ×3 and ×6 facts. Another powerful relationship we want to be sure to explore, however, is that 9 groups are 1 group less than 10 groups. Once again, we want students to determine which facts help them derive the ones they are figuring out. The ×10 facts are relatively easy for many students, so when we can relate ×9 facts, which typically look tricky to students, they may not look tricky anymore.

In the future, understanding ×9 as one group less than ×10 can graduate to ×19 as one group less than ×20, ×99 as one group less than ×100, and so on. Once again, we are not only helping students master their basic facts but unlocking the door to multiplying larger numbers.

WHOLE-CLASS ACTIVITIES

Routines—Math Flips

Berkeley Everett has an amazing collection of cards (including Google Slides versions) that include a variety of images that encourage discussion of number relationships (https:// mathvisuals.wordpress.com/math-flips/). There are 2 sides to the cards, and side A can always be used to solve side B (see Figure 3.61).

Ask the students to determine how many dots are being shown and to explain their thinking. They may skip count, use partial products, or think about 10 groups of 6 being 60.

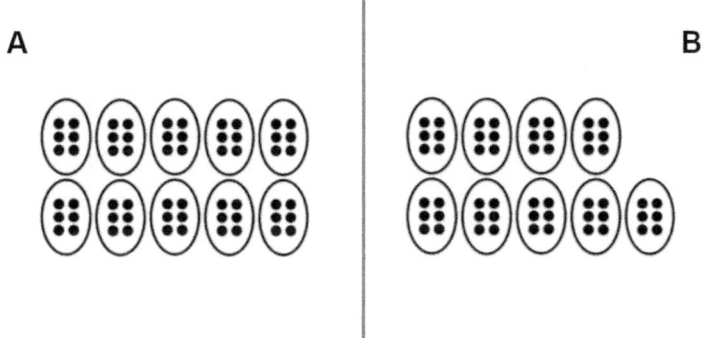

FIGURE 3.61   Math Flip

Prompt: "How does knowing this amount on Side A is 60 help you solve side B?" (Berkeley Everett, 2022)

Discussion can focus on deriving ×9 facts by using ×10 facts.

WHOLE-CLASS MINI LESSON

Introduction: *Teacher gathers students in a meeting area of the room.*

Teacher: *I will be showing you an image, and I want you to determine how many dots there are. Then, we will share how you saw them.*

Teacher then holds up the following image like a sunrise to sunset, giving students a chance to see the image but not enough time for them to count the dots one by one (see Figure 3.62).

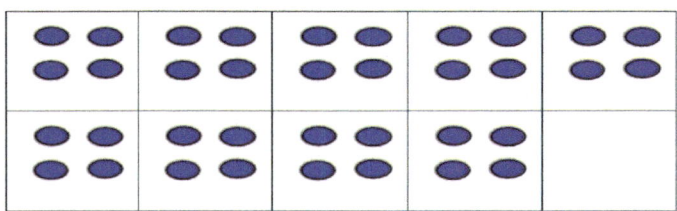

FIGURE 3.62   10 Frame ×4

Teacher: *Put a thumb in front of your chest when you think you know how many dots there are and how you saw them. Once you have one way, try thinking of it another way and then put up another finger to let me know you have two ways of seeing it.*

Jarad: *I think there are 36 of them. I saw them as 5 groups of 4 on the top row and I know that is 20. Then, the bottom row is 4 groups of 4, so I know that is 16. So, when I add 20 plus 16, that's 36.*

Teacher: *Thank you, Jarad. Does anyone have a different way they saw it?*

Jordan: *I saw the 8 groups of 4 and I know that is 32, so I just added one more 4 to get to the 36.*

Teacher: *Thank you, Jordan. Did anyone see it another way?*

Frank: *If the 10 frame were full, that would be 10 groups of 4, which I know is 40. So, there is one 4 less than that for the 9 groups, and 40–4 equals 36.*

Kaitlin: *I saw the top row had 5 groups of 4, which I know is 20. Then, I skip counted from there: 24, 28, 32, and 36.*

Teacher: *Super! So, you all figured out that there were 36 dots, but you saw them in different ways. Let's try another one and, I want to focus on Frank's strategy for using the ×10 facts to help solve the ×9 facts. Remember you may make a mistake when you are trying his strategy, but that means that your brain is learning (see Figure 3.63).*

Teacher shows the following image like a sunrise and then sunset to give time for students to visualize the amount but not time enough to count the dots one by one.

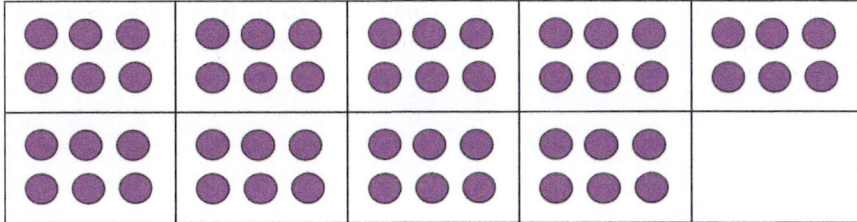

FIGURE 3.63   10 Frame ×6

Adam: I think there are 55 because 10 groups of 6 is 60 and then I counted back 6, 60-59-58-57-56-55.

Teacher: Thank you, Adam. Did anyone see it another way?

Monica: I started with the 60, too, but I know that 6 + 4 is 10, so I know that 54 + 6 = 60.

Adam: That makes sense, but I don't know what I did wrong.

Teacher: If you had 60 and you took away 1, what would you have?

Adam: 59. Oh! So that's what I did wrong. I started with 60. I do know that 6 + 4 makes 10 and that seems to be easier than counting back the 6. So, I think I want to practice that.

Teacher: Yes, using what you know to help you solve what you don't know is a wonderful strategy. Class, what was the focus strategy for today?

Monica: When we are multiplying by a 9, we can use the ×10 facts to help us. You are only one group away.

Teacher: Exactly. Using our ×10 facts to figure out ×9 facts can be a really efficient way of figuring out the answers if you don't know them. You will also probably be more accurate when you are finding your answers since you will not be counting or skip counting. Great thinking today, everyone!

SPOTLIGHT ACTIVITY

If . . . Then . . . (Visual Number Talk)

A powerful instructional strategy is to use sentence stems to help students communicate their thinking and discuss the relationships they are exploring. One daily routine that includes sentence stems is called "If . . . . Then . . .," and focusing on the relationship between ×9 and ×10 it would look like this (see Figure 3.64):

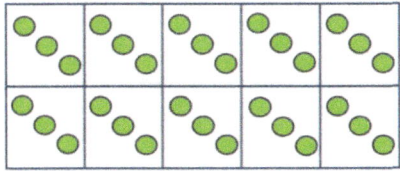

If 10 x 3 = 30....

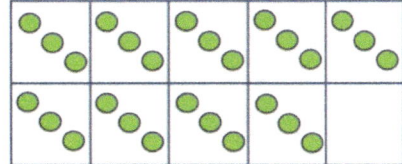

Then 9 x 3 = 27

FIGURE 3.64   If . . . Then . . . Example Dot Pattern

You can build on 10 frames using Cuisenaire® rods so that students are seeing the equal groups existing as a group and won't be as tempted to count them (see Figure 3.65).

If 10 groups of 2 equals _____ ....

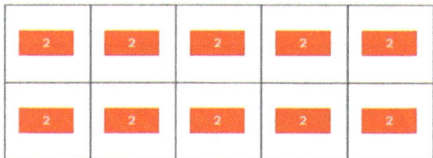

Then 9 groups of 2 equals _____

www.brainingcamp.com

FIGURE 3.65 If . . . Then . . . Cuisenaire® Rods

While the students are building these examples, you can record the matching equations on the board such as $10 \times 2 = 20$ and $9 \times 2 = 18$.

We want to continue to provide explorations concretely, pictorially, and abstractly, along with purposeful practice through daily routines, problem solving, and games, so that students will build their confidence in efficiently using the strategies. As students explore. they are able to use any multiplicative strategy to determine their ×9's, including relating the facts to ×10 but also into any partial products that are known such as ×5 + ×4 (see Figures 3.66–3.74).

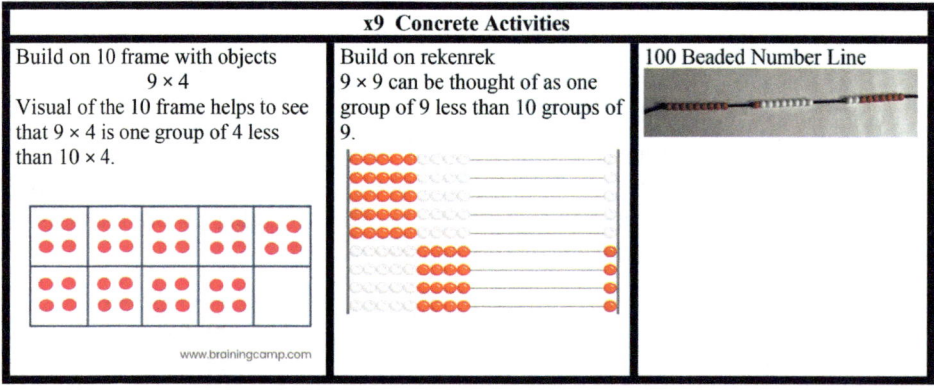

FIGURE 3.66 Concrete Activities

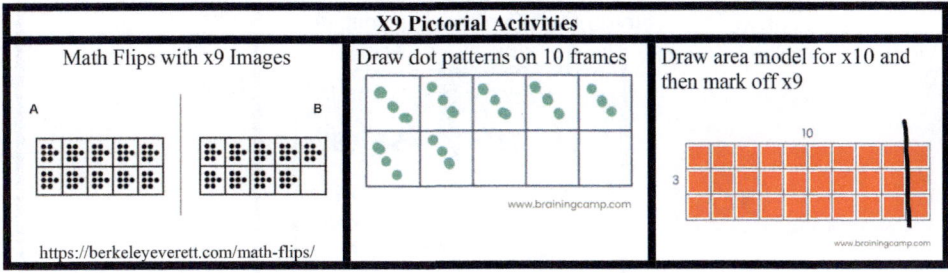

FIGURE 3.67 Pictorial Activities

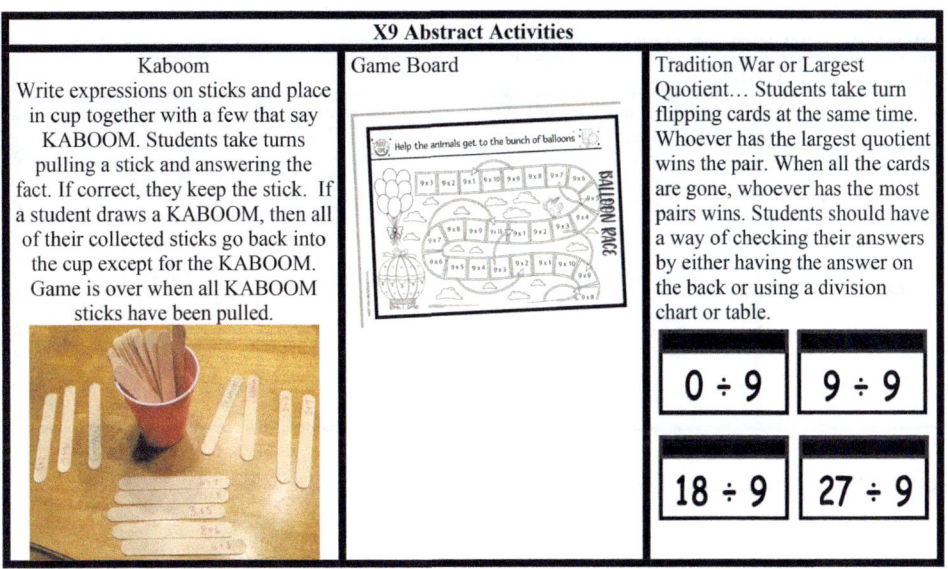

FIGURE 3.68 Abstract Activities

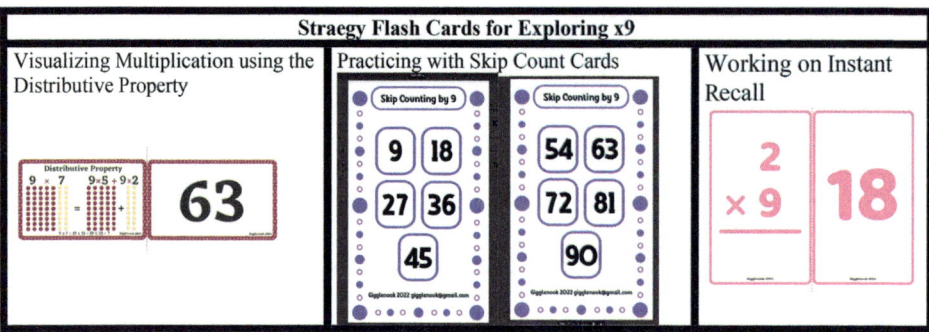

FIGURE 3.69 Strategy Flash Cards

My x9 Story Problems Booklet	Write the set-up equation: Joey was baking 5 pies. He needed 9 apple slices for each pie to decorate the top. How many slices did he need? Show your thinking with a model. Write the solution equation.	Write the set-up equation: Matt raked 9 bags of leaves. Hannah raked 3 times as many bags as Matt. How many bags did Hannah rake? Show your thinking with a model. Write the solution equation.	Write the set-up equation: Josiah played his clarinet song 3 times each day. How many times did he play it in one week? Show your thinking with a model. Write the solution equation.

In every fluency module there should be a focus on word problems. Here are a few examples of the types of word problems that involve x9

FIGURE 3.70 Story Problem Booklet

RESOURCES

Books about x9			
Tang, Greg. The Best of Times (all facts) https://tangmath.com/thebestoftimes	Greg Tang Grapes of Math (all facts) https://tangmath.com/thegrapesofmath		

FIGURE 3.71 Books

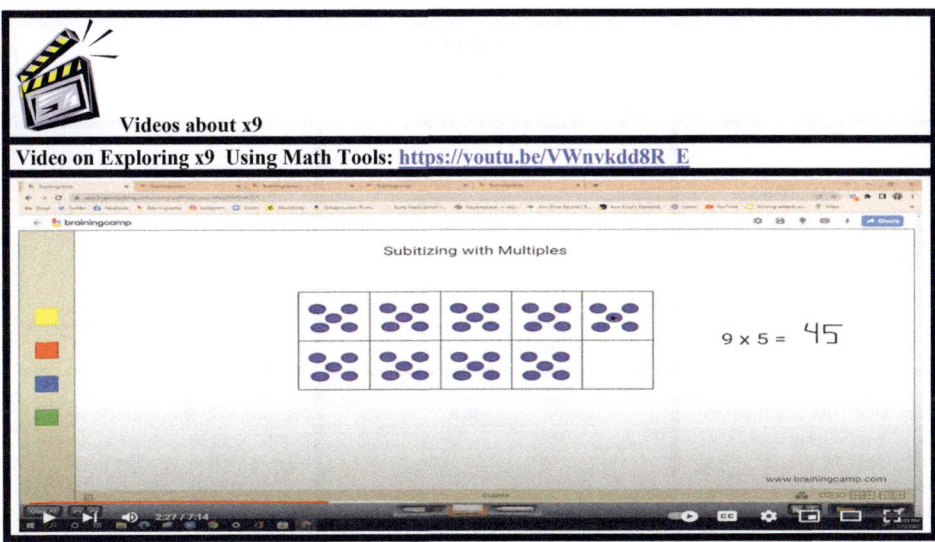

FIGURE 3.72 Video

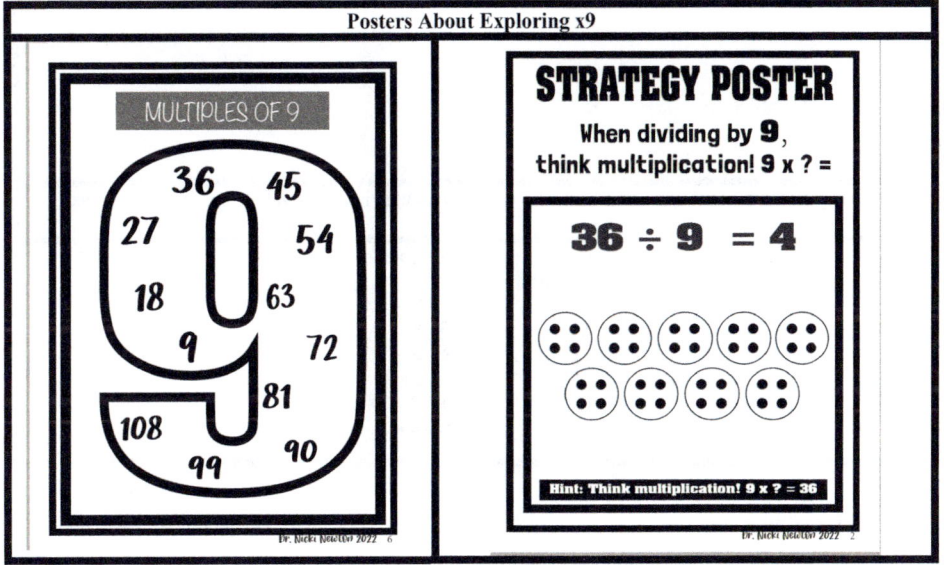

FIGURE 3.73 Posters

Multiplication by 9 Quiz

Name
Date:

2 × 9 = Model with a drawing.	6 × 9 = Model using an area model. 	Solve. Denise needed 9 pieces of ribbon that were each 8 inches long for a project. How many inches of ribbon did she need? Write the equation: Strategy: _____ Answer: _____
Solve. The bakery packed 9 boxes with 8 cookies in each box. They sold 45 cookies. How many cookies do they have left? Write the equation: Strategy: _____ Answer:	Solve. 9 × ____ = 36 _____ × 9= 18 63 = 9 × _____ _____ = 9 × 1	Solve. Jen saw a display that had 9 plants on each of 4 rows. How many plants were there in all? Write the equation: Strategy: _____ Answer: _____

9 × 8 =
Model on the area model in two different ways

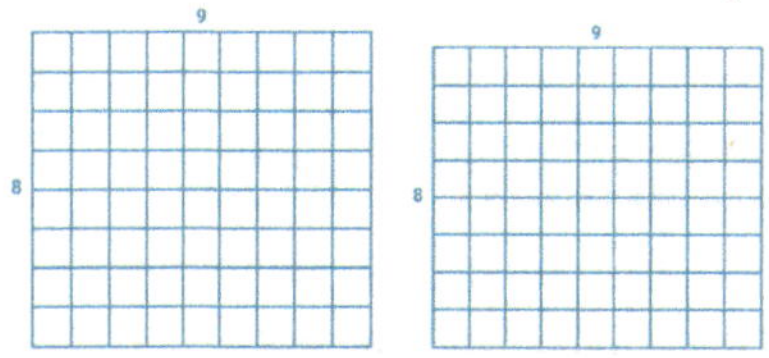

What strategy would you suggest to a friend who doesn't know their 9 facts yet? Explain with numbers, words and pictures.

Circle how good you think you are at doing x9 facts!

Great   Good   Ok, still thinking

FIGURE 3.74   Quiz

EXPLORING AND LEARNING ×7 FACTS

The last set of facts in our suggested progression is the ×7 facts. At this point, if we have mastered all previous sets of facts and highlighted the commutative property of multiplication, the only fact remaining is 7 × 7. We have found in our thousands of student interviews that expressions with 7's are particularly challenging, no matter what the operation is. For multiplication, if we explore 7 groups as the sum of the ×5 and ×2 facts, we can make the ×7 facts much more attainable for all students. This is efficient, unlike when students revert back to skip counting by 7's, which is error prone and time consuming. Taking the time to explore this is worthwhile since some students may prefer this way of partitioning the product over any of the other strategies explored so far. When we provide opportunities for students to develop flexibility with their calculations, we often see the added benefit of increasing their disposition toward math!

WHOLE-CLASS ACTIVITIES

Routines—Model That!

Dr. Nicki Newton's *Daily Math Thinking Routines in Action* features a routine called Model That! Students are given an expression—in this case it would multiplying a.factor by 7—and then model to share how they solved it. Students may model their thinking using concrete objects such as 100-bead rekenreks, 10 frames, Cuisenaire® rods, or number wands. Pictorial models could involve drawing a picture of the groups, an array, or an area model. Abstract representations may include skip counting on a number line or writing equations that demonstrate how the partial products were determined. Students can then do a gallery walk to see all the various strategies, as well as see all the concrete, representational, and abstract ways that thinking can be modeled.

WHOLE-CLASS MINI LESSON

Teacher:	My favorite food is pizza! I could eat it every day. Does anyone else feel that way? I'd like you all to imagine a pizza. How many pieces have you seen pizzas cut into? Class offers a variety of numbers of pieces.
Teacher:	We can cut pizzas into a variety of pieces, but most pizzas I've seen are cut into 8 pieces. For today, I'd like us to think only about pizzas that are cut into 8 pieces. If we had one pizza, how many slices would that be?
Josiah:	8.
Teacher:	Great! Let's record that on this ratio table (see Figure 3.75).

pizzas	slices
1	8

FIGURE 3.75 Ratio Table

Teacher: *Now what if we had 2 pizzas? How many slices would that be?*
Mary: *16.*
Teacher: *How did you figure that out?*
Mary: *I know 8 + 8 = 16.*
Teacher: *Is there another way we can say that that uses multiplication?*
Mary: *2 × 8 = 16.*
Teacher: *Thank you, Mary. I'll record that. I'll also add, what if we had 10 pizzas? I want everyone to think and then put up your thumb in front of your chest when you are ready to share the number of slices as well as how you figured it out (see Figure 3.76).*

pizzas	slices
1	8
2	16
10	

FIGURE 3.76 Ratio Table

Leland: *I think we would have 80 slices because 10 × 8 is the same as 8 tens, and that's 80.*
Teacher: *Does everyone agree? So, I'll add that to the chart, and let's think about how many slices we would have if we had 5 pizzas. See if there is anything we have determined so far that might help you (see Figure 3.77).*

pizzas	slices
1	8
2	16
10	80
5	

FIGURE 3.77 Ratio Table

Delia: We would have 40 slices of pizza because when I think about my 5 facts, I skip count by 5's. So, I did 5, 10, 15, 20, 25, 30, 35, 40.

Teacher: Thanks, Delia. Did anyone think of it another way?

Chris: I got 40, too, but I think of 5 groups as being half of 10 groups. So, 8 × 10 = 80, and half of 80 is 40.

Teacher: Thank you, Chris. The relationship between 5 groups and 10 groups is really powerful. I'll record that and then I'd like you all to think about how many slices we would have if we had 7 pizzas. Remember that you can look at some of the ones we have already done and see if they may help you (see Figure 3.78).

pizzas	slices
1	8
2	16
10	80
5	40
7	

FIGURE 3.78   Ratio Table

Anne: I think we would have 56 slices because 7 groups would be the same as 5 groups and 2 groups added together. So, the 5 × 8 = 40 and 2 × 8 = 16. 40 + 16 = 56.

Teacher: Thank, you Anne. Do you agree with Anne that 7 groups can be thought of as 5 groups and 2 groups added together?

Mary: Yes, when we use the 10-row rekenrek, you can see the colors change at the 5 groups.

Teacher: Could you show us that, Mary?

Mary shows 8 rows of 7 beads on the classroom-sized rekenrek (see Figure 3.79).

FIGURE 3.79   Rekenrek

Delia: *8 × 7 doesn't look as scary once you know you can break the 7 apart into 5 groups and 2 groups.*

Teacher: *Exactly, Delia. We can always use what we know to help us solve what we don't yet know.*

 SPOTLIGHT ACTIVITY

Karate Chop—Distributive Property Exploration

Provide each student with 28 two-sided counters and have them build a 4-by-7 array with all the same color counter. Then, explain that we are going to split this array into two pieces with a karate chop! Model for the students doing the karate chop and then turning over the two-sided counters to change the color for one side of the rectangle, and then represent this with an area model on the board. Then, have students karate chop their arrays and discuss the dimensions of the two partial products (see Figure 3.80). Students can then draw an area model to match. Students can then reset their rectangle back to all the same color again and repeat the activity a few times to find several different solutions when karate chopping the array into two smaller arrays, either vertically or horizontally.

FIGURE 3.80 Color Tile Arrays

One of our favorite parts of the Math Running Record is the math disposition questions because we can see the progress in students' dispositions over the course of the year. One of the questions is to share any expressions from part 1 that look tricky. Inevitably, no matter what the operation, students will mention expressions with a 7 in them. Things look tricky when they are unfamiliar. Once students explore how they can break apart 7 groups into the more familiar ×5 and ×2. Those expressions that once looked tricky no longer look tricky. In the next section are some ideas to provide continued practice opportunities for your students both inside and outside your classroom (see Figures 3.81–3.89).

MATH WORKSTATIONS

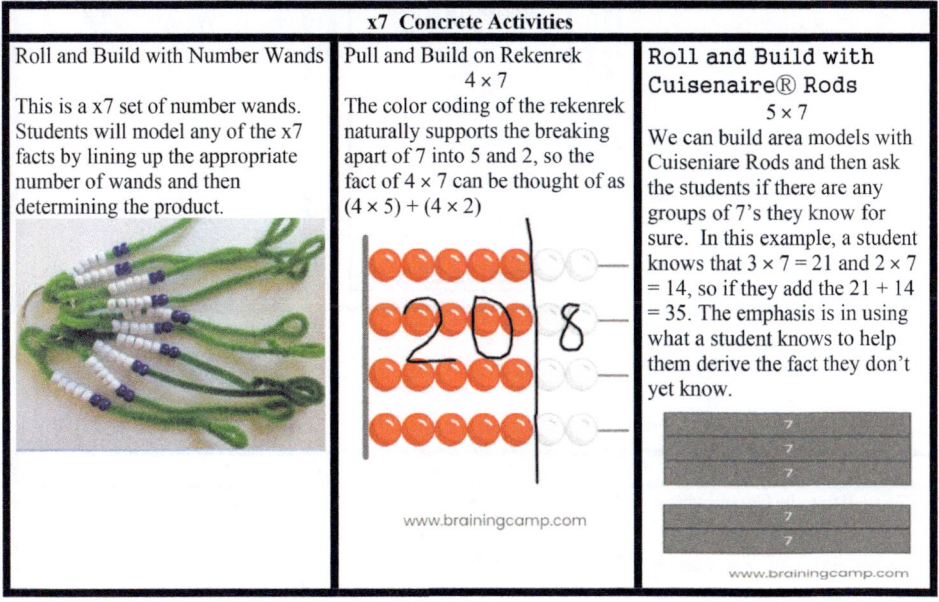

x7 Concrete Activities		
Roll and Build with Number Wands This is a x7 set of number wands. Students will model any of the x7 facts by lining up the appropriate number of wands and then determining the product.	**Pull and Build on Rekenrek** 4 × 7 The color coding of the rekenrek naturally supports the breaking apart of 7 into 5 and 2, so the fact of 4 × 7 can be thought of as (4 × 5) + (4 × 2)	**Roll and Build with Cuisenaire® Rods** 5 × 7 We can build area models with Cuiseniare Rods and then ask the students if there are any groups of 7's they know for sure. In this example, a student knows that 3 × 7 = 21 and 2 × 7 = 14, so if they add the 21 + 14 = 35. The emphasis is in using what a student knows to help them derive the fact they don't yet know.

FIGURE 3.81 Concrete Activities

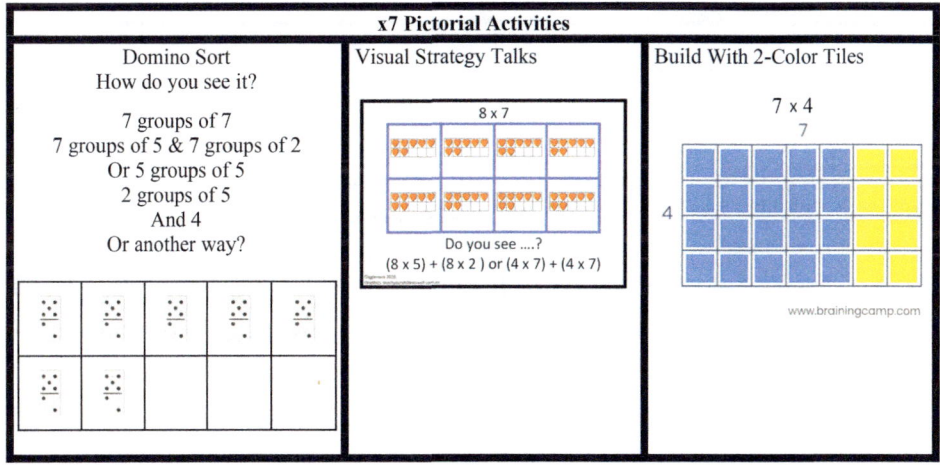

x7 Pictorial Activities		
Domino Sort How do you see it? 7 groups of 7 7 groups of 5 & 7 groups of 2 Or 5 groups of 5 2 groups of 5 And 4 Or another way?	**Visual Strategy Talks** 8 × 7 Do you see ....? (8 x 5) + (8 x 2) or (4 x 7) + (4 x 7)	**Build With 2-Color Tiles** 7 × 4 7

FIGURE 3.82 Pictorial Activities

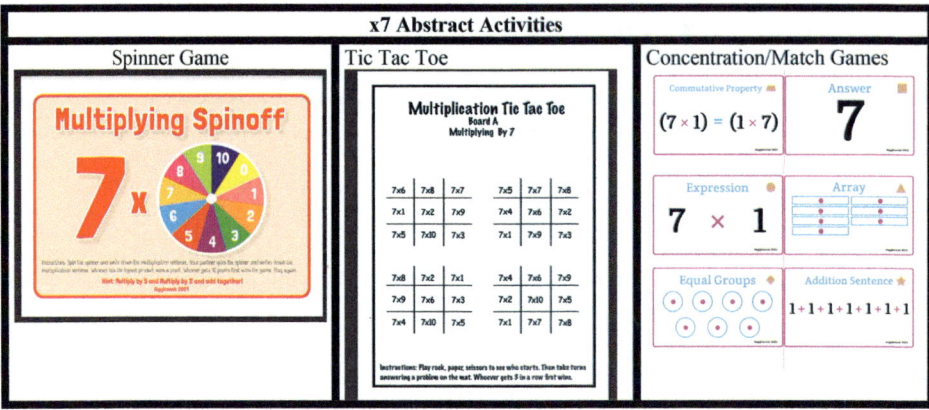

FIGURE 3.83 Abstract Activities

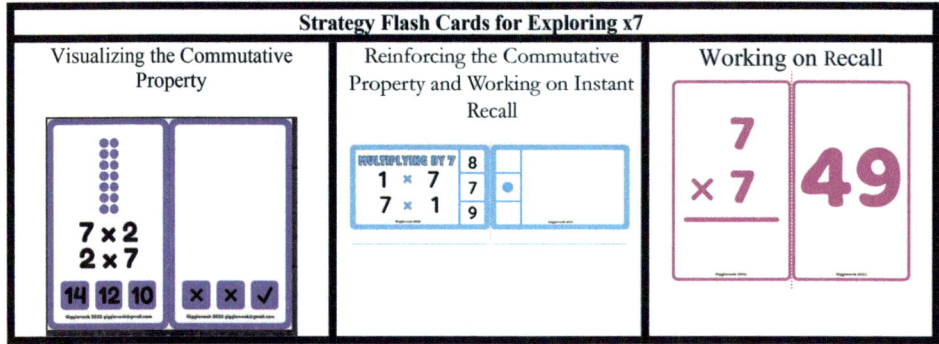

FIGURE 3.84 Strategy Flash Cards

In every fluency module there should be a focus on word problems. Here are a few examples of the types of word problems that involve x7			
My x7 Story Problems Booklet	**Write the set-up equation:** Outside a store there were 7 bicycles. How many wheels were there?	**Write the set-up equation:** An apartment building had 7 floors. If there were 6 apartments on each floor, how many apartments were there?	**Write the set-up equation:** At the pet store, there was an aquarium with 4 guppy fish. The tank next to it had 7 times as many neon fish as the guppy tank. How many neon fish were there?
	Show your thinking with a model.	**Show your thinking with a model.**	**Show your thinking with a model.**
	Write the solution equation.	**Write the solution equation.**	**Write the solution equation.**

FIGURE 3.85 Story Problem Booklets

RESOURCES

Books about x7			
Pam Calvert Multiplying Menace The Revenge of Rumplestilskin (all facts, includes multiplying by a fraction) https://www.youtube.com/watch?v=53XH9ASKatM	Greg Tang Grapes of Math (all facts) https://tangmath.com/thegrapesofmath	Greg Tang Best of Times (all facts) https://tangmath.com/thebestoftimes	Elinor Princzes My Full Moon is Square (square numbers) https://www.youtube.com/watch?v=K5Uzv2C8In0
Hulme, Joy N. Sea Squares https://www.youtube.com/watch?v=tFumnNXLzmE			

FIGURE 3.86 Books

Videos about x7

Video on Exploring x7 Using Math Tools: https://youtu.be/JXvBm2UvI_E

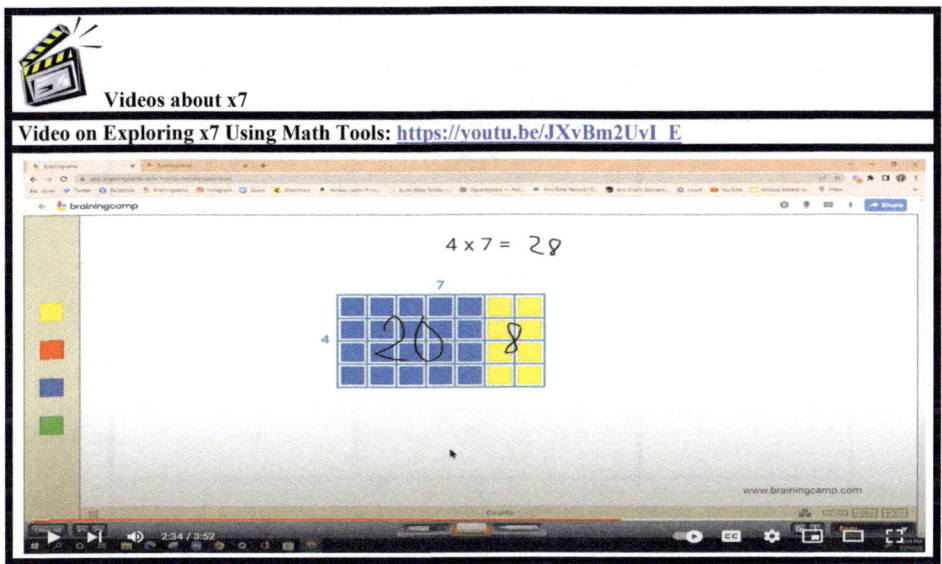

FIGURE 3.87 Video

FIGURE 3.88 Posters

Multiply by 7 Quiz

Name
Date:

$7 \times 3 =$ Model with a drawing.	$6 \times 7 =$ Model on the area model. 	Solve. During physical education class, students were put onto teams that had 7 students in each. If there were 4 teams, how many students were there in the class? Write the equation: Strategy: _____ Answer: _____
Solve. On one sheet of stickers, there were 7 rows each having 9 stickers. How many stickers were on the sheet? Write the equation: Strategy: _____ Answer: _____	Solve. $7 \times$ ____ $= 35$ ____ $\times 7 = 0$ $49 = 7 \times$ _____ ____ $= 7 \times 7$	Solve. Dan ran 2 miles every day. How many miles did he run in a week? Write the equation: Strategy: _____ Answer: _____

$7 \times 6 =$
Model on the area model in two different ways.

What strategy do you use when you multiply by a 7? Explain with numbers, words and pictures.

Circle how good you think you are at doing x7 facts!
Great   Good   Ok, still thinking

FIGURE 3.89 Quiz

KEY POINTS

- Learning the trajectory of multiplication
- Foundational facts
- Derived facts
- Goal setting

SUMMARY

Teaching multiplication is complex. There are foundational facts that are meant to be used as the foundation for multiplicative thinking with the other facts. Multiplication is a journey. It is important to make sure that we always consider where students are on the journey and then, based on that information, decide what comes next. It is important to have students be actively involved in the journey, reflecting on what they know, setting goals, and celebrating their successes. Every child can learn to multiply!

REFLECTION QUESTIONS

1. What stands out for you in this chapter?
2. What are your strengths when teaching multiplication, and what are your challenges? What new ideas do you have?
3. Can you state or draw a three-step action plan based on this chapter?

CALL TO ACTION

 1. Share your favorite "Aha!" moment about teaching multiplication on social media to help spread the movement! #FDJH

 2. Take a photo of different ways that you are modeling multiplication math facts in your classroom and share it on social media to encourage other teachers to do it, too! #FDJH

3. Get started with reasoning about multiplication with this balance puzzle card game!

REFERENCES

Baroody, A. (2006, August). Why children have difficulties mastering the basic number combinations and how to help them. *Teaching Children Mathematics, 13*(1), 22–23.

Bay-Williams, J., & SanGiovanni, J. (2021). *Figuring out fluency in mathematics teaching and learning, grades K-8: Moving beyond basic facts and memorization.* Corwin.

Boaler, J., Williams, C., & Confer, A. (2015). Fluency without fear: Research Evidence on the best ways to learn math facts. *YouCubed.* www.youcubed.org/evidence/fluency-without-fear.

Brendefur, J., Strother, S., Thiede, K., & Appleton, S. (2015, August). Developing multiplication fact fluency. *Advances in Social Sciences Research Journal, 2*(8), 142–154.

Everett, B. (2022). https://berkeleyeverett.com/math-flips/

Fosnot, C. T., & Dolk, M. (2001). *Young mathematicians at work: Constructing number sense, addition and subtraction.* Heinemann.

Karp, K. S., Bush, S. B., & Dougherty, B. J. (2014, August). 13 rules that expire. *Teaching Children Mathematics, 21*(1), 18–25.

Kling, G., & Bay-Williams, J. M. (2021, November). Eight unproductive practices in developing fact fluency. *Mathematics Teacher: Learning and Teaching PK-12, 114*(11), 830–838.

Looney, S. (2022). https://www.samebutdifferentmath.com/multiplication-division

National Council of Teachers of Mathematics (NCTM). (2014). *Principles to actions: Ensuring mathematical success for all.* NCTM.

National Research Council, Mathematics Learning Study Committee. (2001). *Adding it up: Helping children learn mathematics* (Kilpatrick, J., Swafford, J., & Findell, B., Eds.). The National Academies Press.

Newton, N. (2016). *Math running records in action: A framework for assessing basic fact fluency in grades K-5* (1st ed.). Routledge. https://doi.org/10.4324/9781315682389

Newton, N. (2018). *Daily math thinking routines in action: Distributed practices across the year* (1st ed.). Routledge. https://doi.org/10.4324/9781351164283

Sherin, B., & Fuson, K. (2005). Multiplication strategies and the appropriation of computational resources. *Journal for Research in Mathematics Education, 36*(4), 347–395.

Van de Walle, J. A., Karp, K. S., & Bay Williams, J. M. (2013). *Elementary and middle school mathematics: Teaching developmentally* (8th ed.). Pearson.

CHAPTER 4

Exploring and Learning Division Facts

Ann Elise has been known to say that subtraction is the Achilles heel of our students, but that's only because division is always late to the party! Typically, in state standards, we have three years (K–2) to help our students achieve mastery with understanding with addition and subtraction. For multiplication and division, however, we typically are given just one year, and that's for both! Since much of the year is focused on the conceptual understanding of the operations and the mastery of multiplication math facts, little time tends to be devoted to division. Research has shown that the thinking students do when asked division facts relies heavily on them thinking about their multiplication facts (Kouba, 1989; Mulligan & Mitchelmore, 1997; National Research Council, Mathematics Learning Study Committee, 2001), so dedicating time to developing mastery of multiplication facts will positively affect students' mastery of their division facts. The key is making time for this. While there aren't as many strategies for division as there are for multiplication, there are some specific challenges and models we want to explore that will help students form a foundation for understanding division and thinking and reasoning multiplicatively about division that students may generalize across larger numbers down the road.

The use of story contexts when we begin exploring division, and our modeling of those structures, is a necessary beginning step for our students. We will go into more detail in Chapter 5 on ways we explore story problems that involve division. For now, let's explore the two types of equal group problems (Carpenter et al., 2014): partitive and quotative. In both cases, we know the total amount of objects. In partitive division situations, we know the *number of groups*, so we need to determine the fair share amount in each group. In quotative division scenarios, we know the *amount in each group*, and we need to determine how many groups there are in the dividend.

As students begin to explore these contexts, they go through a progression of thinking beginning with direct modeling where they act out the story situation with objects or fingers, moving to skip counting either forward or backward, and then using derived facts where they think and reason, typically using the multiplication facts they know to help them derive the division fact they are determining. With multiplication, there are a variety of relationships and strategies that students can use to find products they are unable to retrieve them from memory. While there is less research related to how students think when faced with division problems, the available research suggests that students predominantly think multiplication when they are asked to solve a division fact (Carpenter et al., 2014; Kouba, 1989; Mulligan & Mitchelmore, 1997). For example, when a student is

DOI: 10.4324/9781032614229-4

asked 48 ÷ 8, they tend to think 8 times something is 48 on their way to determining the missing factor is 6.

It is crucial that we talk with our students about their thinking. When using Dr. Nicki Newton's Math Running Records (www.mathrunningrecords.com), we can determine which set of facts students need to begin working with as well as where they are with the strategy levels. This information helps us to best facilitate their progress along the continuum. When administering this diagnostic assessment, we must carefully watch how students engage with each fact. Do they produce it automatically—or as one student once described, it as the number "popping like popcorn in my mind?"

Of particular interest is when a student is unable to retrieve an answer from memory when asked to solve for a specific fact. What we need to look for is what happens next and how they proceed. Some students fall back in the progression of strategy levels and count one by one to determine groups; others skip count forward or backward. The difference in counting is critical because it reveals *how* students are thinking. We want to help our students think and reason *multiplicatively* instead of *additively*. This will support efficiency when determining quotients with basic facts and can be applied to all sets of numbers they will encounter. In this chapter, we will explore these strategies.

Once again, you will notice that the sequence of the facts we suggest does not follow the sequential order of the factors. Indeed, teaching the facts in order of factor size is something that Kling and Bay-Williams (2021) name as one of the eight unproductive practices for building fact fluency. Instead, we want to explore strategies based on a number relationships that are easier for our students to understand and build from there (Bay-Williams & SanGiovanni, 2021; Baroody, 2006; Brendefur et al., 2015). There are several sets of facts that we have noticed during our administration of Math Running Records that provide their own thinking challenge, so we will begin with these.

Although we have separate chapters for multiplication and division strategies, we are fascinated by the possibilities of exploring the multiplication and division facts simultaneously with students. Not much research has been done on the impact on exploring multiplication and division facts together, but the Common Core State Standards (CCSS), Texas Essential Knowledge and Skills (TEKS), and National Council of Teachers of Mathematics (NCTM) all mention that students should solve problems within 100 using strategies based on the relationship between multiplication and division, as well as the properties. We suggest that exploring the facts simultaneously would be one way that we could be sure students develop this conceptual understanding of the inverse relationships between the two operations (see Figures 4.1 and 4.2).

TOOLS, TEMPLATES, ACTIVITIES, AND GAMES

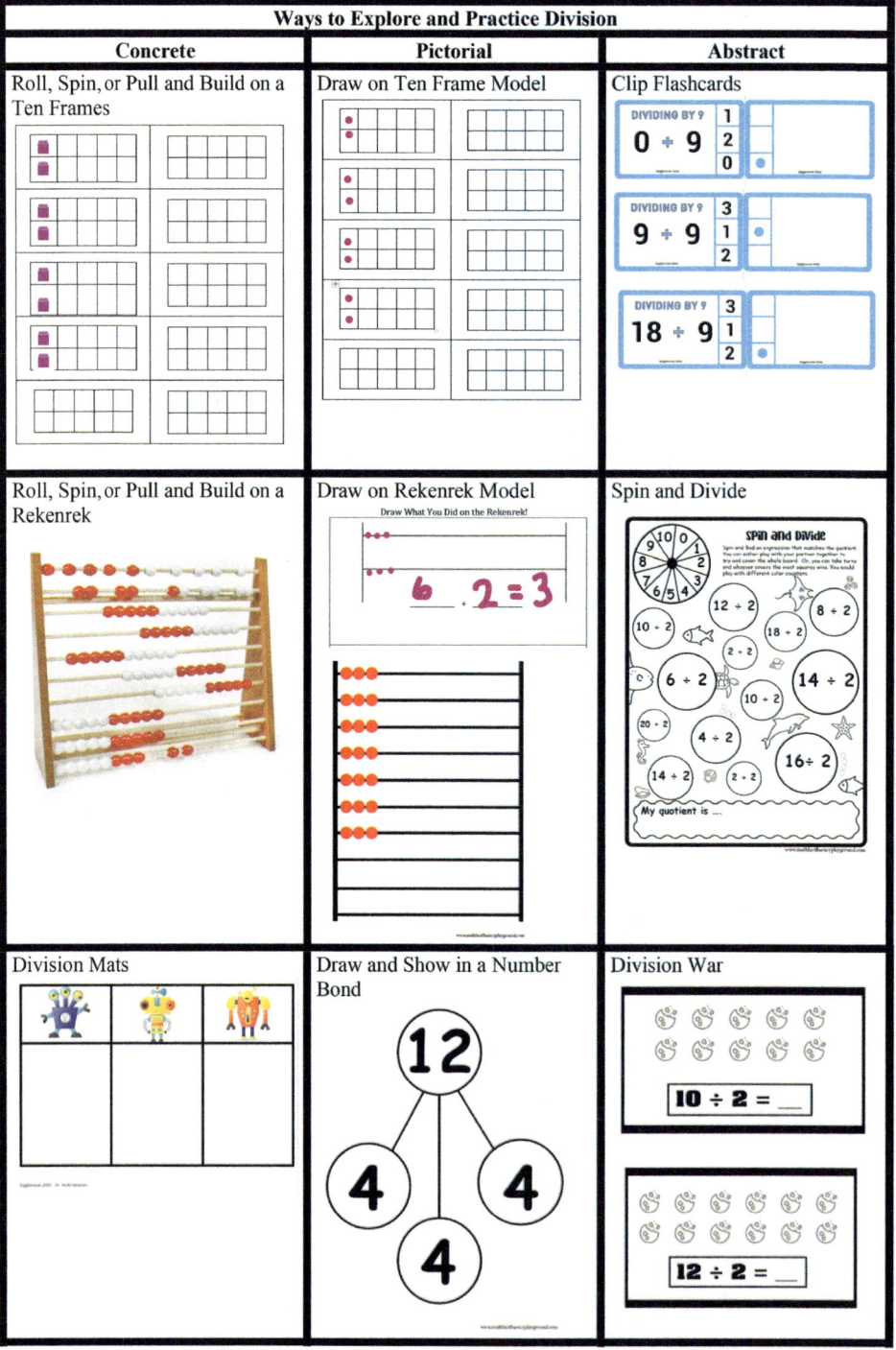

FIGURE 4.1 Explorations

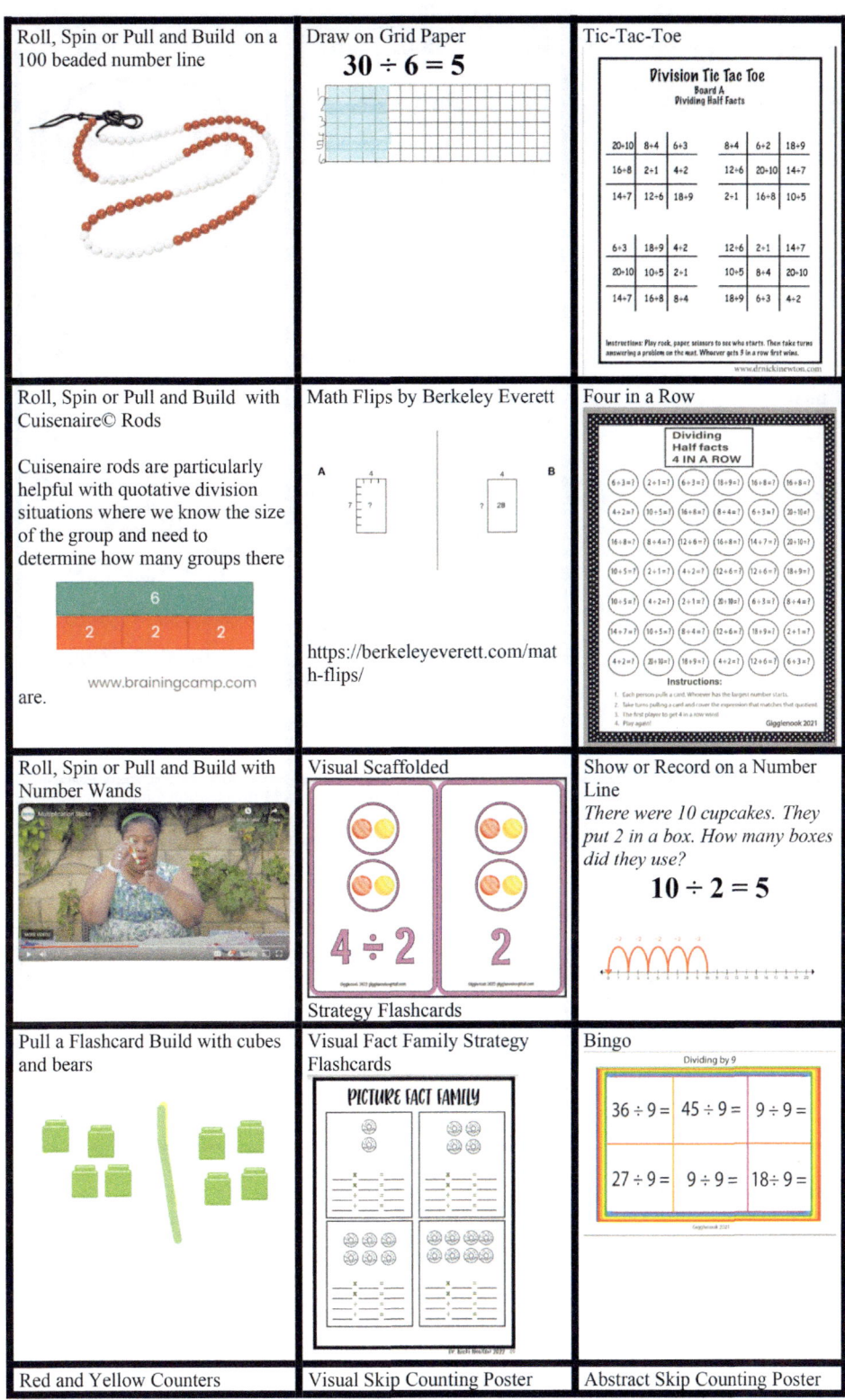

Roll, Spin or Pull and Build on a 100 beaded number line	Draw on Grid Paper $30 \div 6 = 5$	Tic-Tac-Toe
Roll, Spin or Pull and Build with Cuisenaire© Rods. Cuisenaire rods are particularly helpful with quotative division situations where we know the size of the group and need to determine how many groups there are.	Math Flips by Berkeley Everett https://berkeleyeverett.com/math-flips/	Four in a Row
Roll, Spin or Pull and Build with Number Wands	Visual Scaffolded	Show or Record on a Number Line *There were 10 cupcakes. They put 2 in a box. How many boxes did they use?* $10 \div 2 = 5$
Pull a Flashcard Build with cubes and bears	Visual Fact Family Strategy Flashcards	Bingo
Red and Yellow Counters	Visual Skip Counting Poster	Abstract Skip Counting Poster

FIGURE 4.1   (Continued)

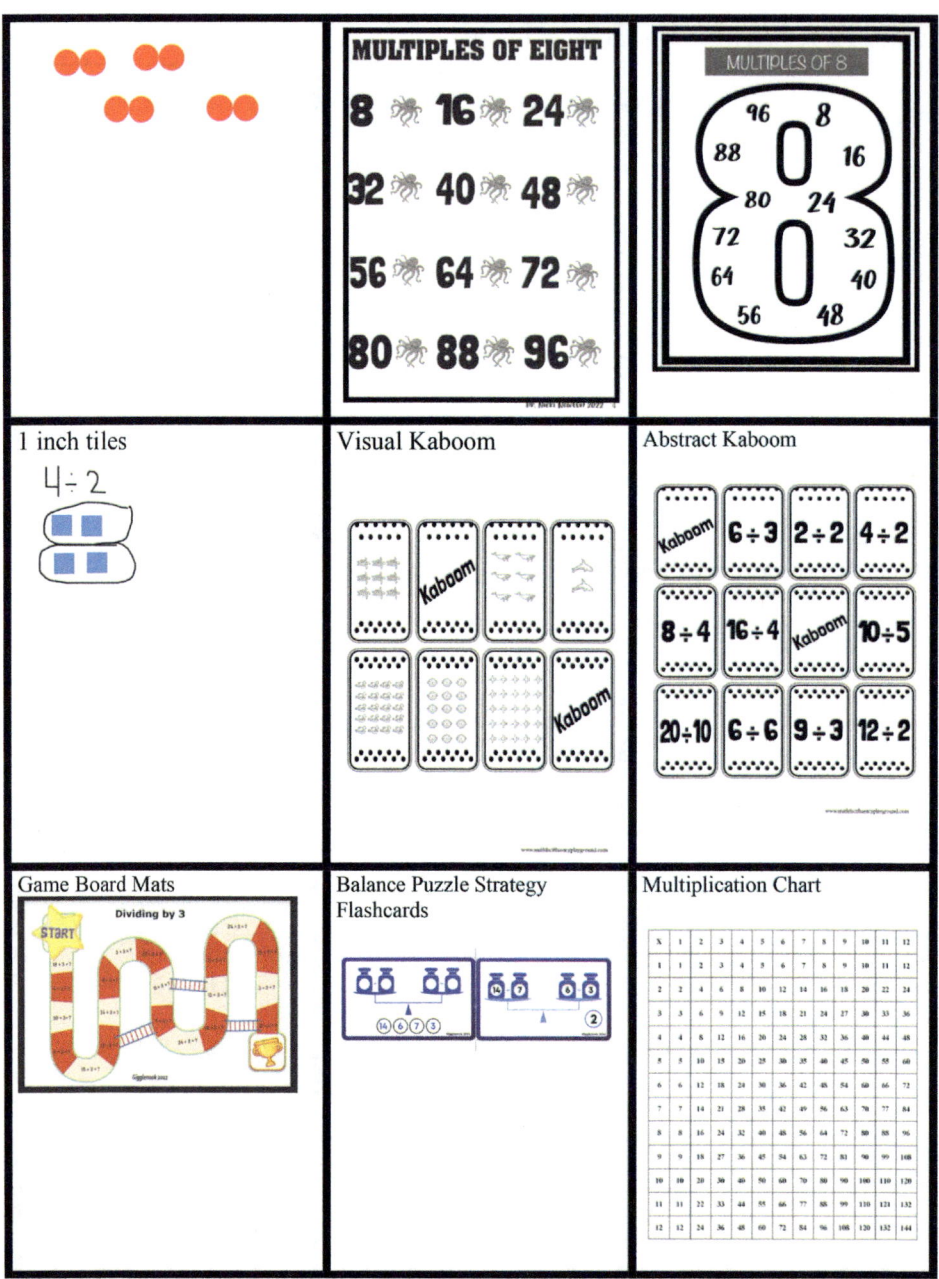

FIGURE 4.1 (Continued)

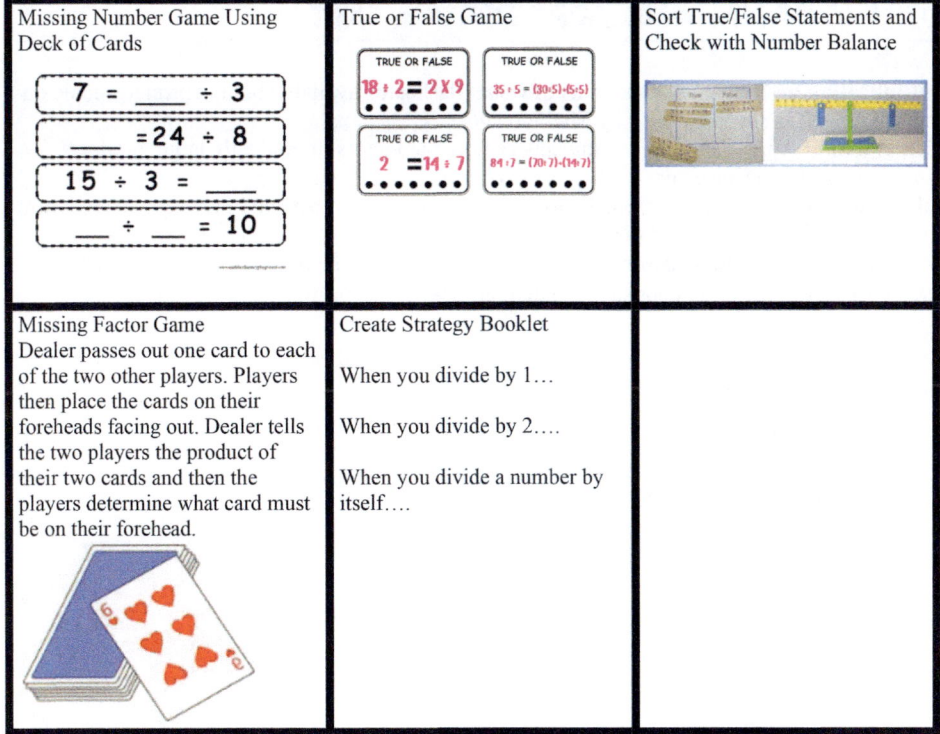

Missing Number Game Using Deck of Cards	True or False Game	Sort True/False Statements and Check with Number Balance
7 = ___ ÷ 3 ___ = 24 ÷ 8 15 ÷ 3 = ___ ___ ÷ ___ = 10	TRUE OR FALSE $18 ÷ 2 = 2 × 9$ TRUE OR FALSE $35 ÷ 5 = (30÷5)+(5÷5)$ TRUE OR FALSE $2 = 14 ÷ 7$ TRUE OR FALSE $84 ÷ 7 = (70÷7)-(14÷7)$	
Missing Factor Game Dealer passes out one card to each of the two other players. Players then place the cards on their foreheads facing out. Dealer tells the two players the product of their two cards and then the players determine what card must be on their forehead.	Create Strategy Booklet When you divide by 1… When you divide by 2…. When you divide a number by itself….	

FIGURE 4.1   (Continued)

ONLINE GAMES AND APPS FOR ALL FACTS

Multiplication by Heart - Fabulous set of visual flashcards available both in print available on Amazon as well as for free online by Mathigon here: https://mathigon.org/multiply
Stick and Split App - This app has students explore the factors of gradually larger products using rods similar to Cuisenaire rods. https://www.maypoleeducation.com
Kakooma - This is a game where students are given several numbers and they need to recognize a fact family of two factors and their product. https://tangmath.com/kakooma
Fast Facts (Don't let the name fool you. You know we don't like timed activities. This app offers visuals to support conceptual understanding.) https://apps.apple.com/us/app/fast-facts-math/id506232953

FIGURE 4.2 Online Games

EXPLORING AND LEARNING ÷1 FACTS

Once students have had the opportunity to develop their conceptual understanding of division, they will then be able to begin developing their mastery of division facts. In our experience, we have found that the first three facts on the progressions—divide by 1, divide 0 by numbers, and divide a number by itself—can be super confusing for many students. Students need to spend time on each of these sets of facts so that they not only achieve accurate answers but understand *why* those are the answers. Too often, when we ask students how they knew that 8 ÷ 1 = 8, they will say, "My teacher told me that anything divided by 1 is itself." So rather than truly understanding why 8 is the answer, students have learned a rule that merely helps them get the correct answer. Students need to understand that when dividing by 1, we are determining h*ow many groups of 1 are in the dividend*, or the amount there will be if one group is formed. As will be true with all the division facts, we can think multiplication. So, when asked 4 ÷ 1, we can think 1 times something must equal 4.

WHOLE-CLASS ACTIVITIES

Routines—Why Is It Not?

In Dr. Nicki Newton's *Daily Math Thinking Routines in Action*, she shares a routine called Why Is It Not?! In this routine, students need to reason about why a given answer can't be the correct answer and then prove their thinking. An example might be, "8 ÷ 1 = _____. Why is the answer not 1?" Students are then encouraged to use any concrete objects or pictures to support their thinking to their classmates.

WHOLE-CLASS MINI LESSON: DIVIDING BY 1

Introduction: *Teacher ensures each student has 6 linking cubes and a mat.*

Teacher:	*I'm going to tell you all a story, and I'd like you to use the cubes to act out the situation. Here we go. Marty has 6 dog bones. If he digs 1 hole in the yard for each of his bones, how many holes does he need to dig?*
	Students then move the cubes into 6 separate piles.
Teacher:	*Can someone share how many holes Marty will dig and how you know?*
Jaritza:	*He will dig 6 holes because this bone will get a hole (Jaritza places a cube to one side), and then this bone will get a hole (she moves another cube to be by itself, and so on until all 6 cubes are situated by themselves). So, each of the 6 bones will get its own hole.*
Teacher:	*Thank you, Jaritza. Let's act out another story. A baker has boxes that can fit 1 big cupcake. If the baker has 4 cupcakes, how many boxes does she need if they each need to be in a box? Let's act this out with the cubes again.*
	Teacher notices that Sally has taken 4 cubes and placed them scattered in front of her.

Sally: She needs 4 boxes because one cupcake goes in one box, then the second one goes in a box, the third one gets a box, and the fourth one, too. *As Sally discusses each cupcake, she points to each cube on her mat.*

Teacher: Thumbs up if you agree with Sally that the baker would need 4 boxes. *Teacher sees that all students are giving a thumbs up.*

Teacher: Super! He's another story for you. Enrique has 5 books. If he gives 1 book each to his friends, how many friends does he have books for? *Students take 5 cubes and place them in separate piles.*

Antonia: Enrique has 5 books, so he can give 1 each to 5 friends. I'm noticing something. Each story you have told us has an answer of the number we are starting with.

Teacher: I wonder if that will always be true?

Antonia: Well, it has to, doesn't it? With the bones, we had 6 of them and they each needed 1 hole. So there needed to be 6 holes. No matter how many bones there are, we would need that many holes.

Martin: And for the cupcakes, each box only fit 1 cupcake, and so we needed the same number of boxes as we had cupcakes.

Teacher: I love your observations! How do you think we could represent these situations using a math equation?

Sally: You are dividing by 1, so it could be $6 \div 1 = 6$ for the bones story.

Jeff: For the cupcakes, it could be $4 \div 1 = 4$.

Sandy: And for the books, it would be $5 \div 1 = 5$.

Teacher: I think you have noticed a really important relationship that will help us when we are dividing any number by 1. Let's add this to our anchor chart. *Teacher writes "We suggest that anything divided by 1 will be itself" onto the anchor chart.*

SPOTLIGHT ACTIVITY

Make a play bakery and pack things. Use boxes, plates, or even just paper bags. Act out stories.

> Story 1: The bakery had 3 cupcakes. They put all of them in 1 box. How many cupcakes did the bakery have in a box?
>
> Story 2: The bakery had 8 cupcakes. They put 1 cupcake in each box. How many boxes did they use?

Continue having the students act out these stories in front of the class and also on work mats (see Figures 4.3–4.5). It is essential that you share both partitive and quotative stories throughout your unit on division and throughout the year.

FIGURE 4.3 Bakery Boxes

FIGURE 4.4 AND 4.5 Bakery Boxes and Donuts

It is important to teach division throughout the year and not just wait until you complete your study of multiplication. The initial unit of study is a deep exploration, but the learning takes place across the year as students engage in continued practice with the concepts through games, energizers, routines, and problem solving (see Figures 4.6–4.13).

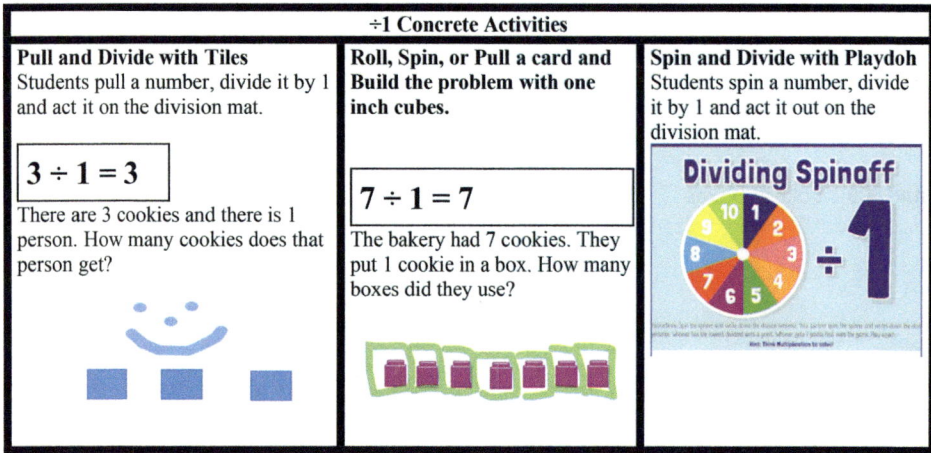

FIGURE 4.6 Concrete Activities

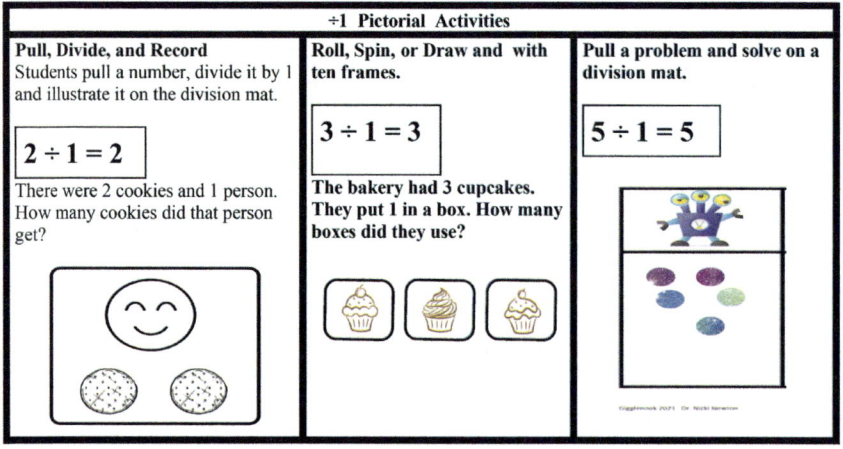

FIGURE 4.7 Pictorial Activities

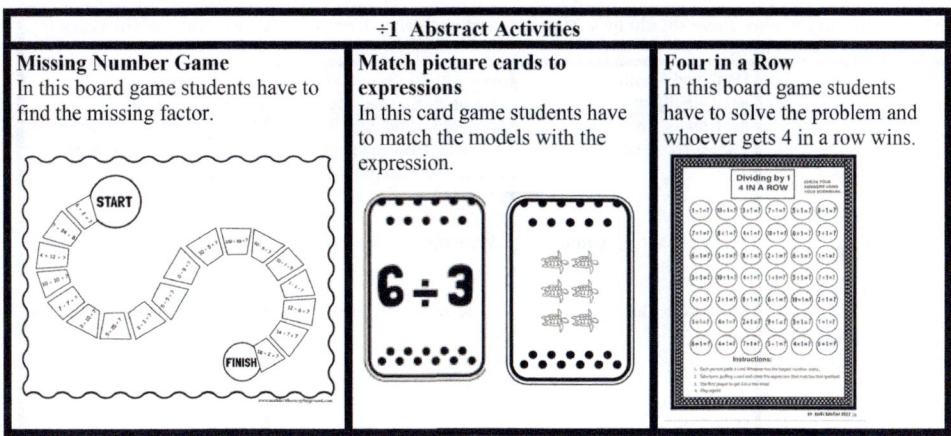

FIGURE 4.8 Abstract Activities

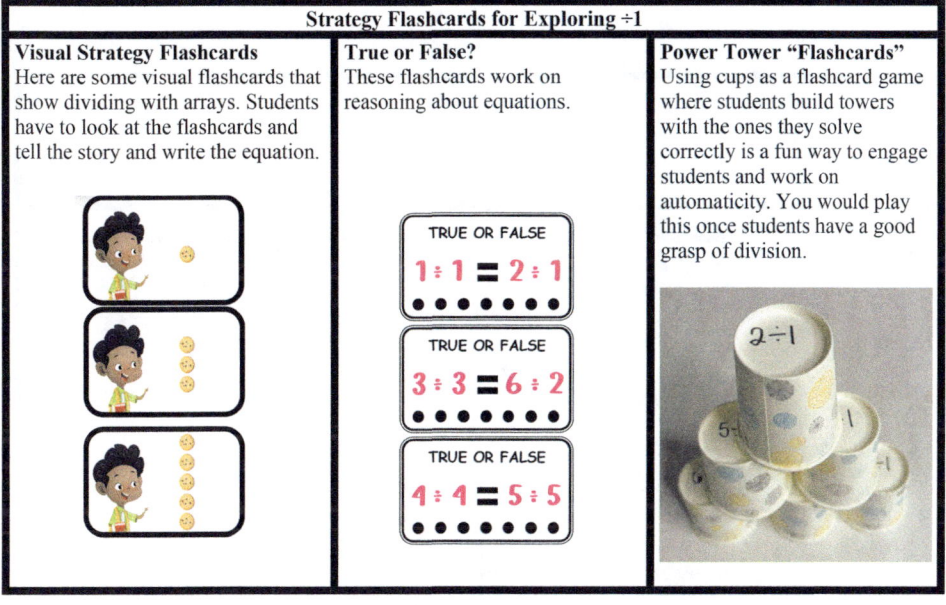

FIGURE 4.9 Strategy Flashcards

In every fluency module there should be a focus on word problems. Here are a few examples of the types of word problems that involve ÷1			
My ÷1 Story Problems Booklet	Hank had 5 balloons. If he gave 1 balloon to each of his friends, how many friends received a balloon?	Aiden had 5 decorated cupcakes. She had special bags that could only fit 1 cupcake. How many bags did she need to put them all in a bag?	There were 5 beads arranged in rows. If they were arranged in one row, how many rows of beads were there?
	Write the set-up equation:	Write the set-up equation:	Write the set-up equation:
	Show your thinking with a model.	Show your thinking with a model.	Show your thinking with a model.
	Write the solution equation.	Write the solution equation.	Write the solution equation.

FIGURE 4.10 Story Problem Booklet

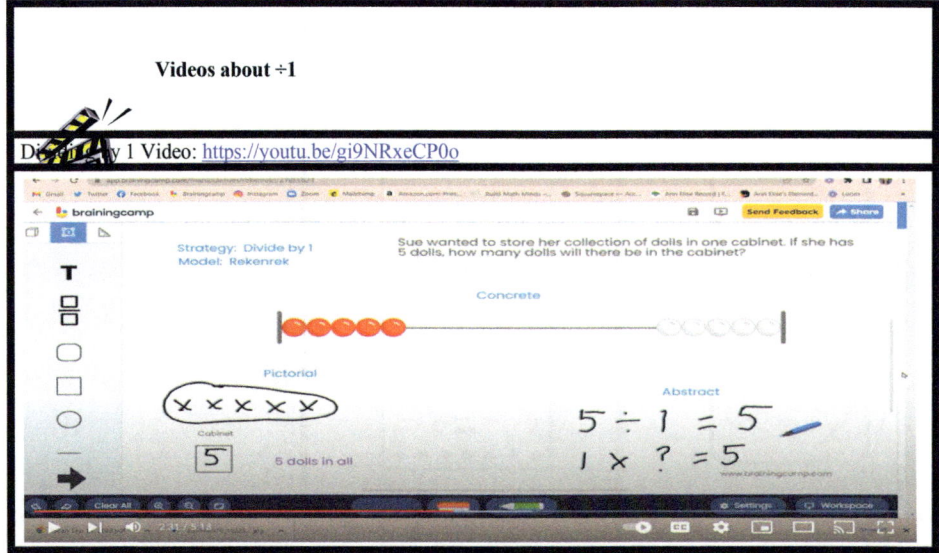

Videos about ÷1

Divide by 1 Video: https://youtu.be/gi9NRxeCP0o

FIGURE 4.11 Resources

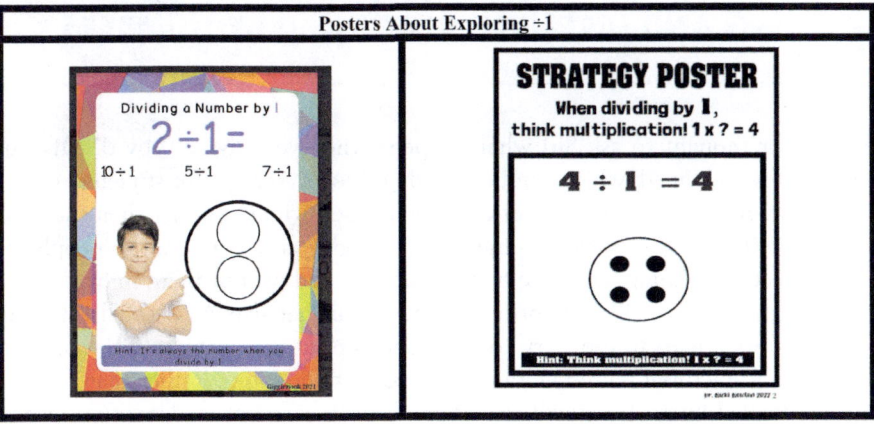

FIGURE 4.12 Posters

Dividing by 1 Quiz		
Name: Date:		
$3 \div 1 =$ Model using equal groups.	$6 \div 1 =$ Model using an array.	Solve. There were 7 runners on a racetrack. If each runner had their own lane, how many lanes did they use? Write the equation: Strategy: _____ Answer: _____
Solve. There were 9 display windows at a Lego store. If each display had one structure, how many structures were there? Write the equation: Strategy: _____ Answer:	Solve. $6 \div \underline{} = 6$ $\underline{} \div 1 = 3$ $8 = \underline{} \div 1$ $\underline{} = 7 \div 1$	What is the $\div 1$ strategy? Explain with numbers, words, or pictures
$6 \div 1 =$ Model on the number line.		
What happens when we divide by 1? Explain with numbers, words and pictures.		
Circle how good you think you are at doing $\div 0$ and $\div 1$ facts! Great Good Ok, still thinking		

FIGURE 4.13 Dividing by 1 Quiz

EXPLORING AND LEARNING DIVIDING 0 BY NUMBERS FACTS

Have you ever thought to ask Siri what happens when you divide 0 by 0? Of course, dividing by 0 is undefined, which can be a really difficult concept for students to understand and reason about. For this set of facts, then, the dividend, rather than the divisor, will always be 0. When we have no objects and we give an equal share to people, each person will receive nothing. Or, if we have no objects and we put them in bags that can fit a certain amount, we will still not have any bags of that size. As with all concepts, it is important for students to use story contexts and act out the contexts with objects and connect them to pictorial representations and abstract equations. Rather than learning a rule that 0 divided by anything will always result in 0, we want students to understand why that is. When asked $0 \div 4$, students can think something times 4 equals 0.

WHOLE-CLASS ACTIVITIES

Routines—True or False

In Dr. Nicki Newton's *Daily Math Thinking Routines in Action*, she shares a routine called True or False. In this routine, students will be given an expression and asked to determine whether it is true or false. They may then partner share and use concrete objects or pictures to support their thinking about whether the statement is true or false. Some examples are "$0 \div 4 = 4$," "$0 \div 6 = 0 \times 6$," or, "When you divide 0 by numbers, the answer will always be 0."

WHOLE-CLASS MINI LESSON: DIVIDING 0 BY NUMBERS

Teacher ensures each student has some linking cubes and a mat.

Teacher:	*Today we are going to talk about dividing 0 by numbers. Let me give you a story that can help us begin to think about this. Shanique has 4 cookies. If she shares them with a friend, how many cookies can each person get?*
Lorraine:	*(Pointing to her cubes, which are in two groups of two) Each person would get 2 cookies. The 2 groups of 2 make 4 in all.*
Teacher:	*Thumbs up if you agree with Lorraine.*
	Teacher notices that all students have their thumbs up.
Teacher:	*Super! How would we write that as a number sentence?*
Joey:	*I think it would be $4 \div 2 = 2$ because you had the 4 cookies and then shared them between 2 people. Each person got 2.*
Teacher:	*Is there a way we can think about it that would help us double check?*
Jessica:	*We can think about multiplication. $2 \times 2 = 4$, so it has to be true that $4 \div 2 = 2$.*

Teacher:	*Yes, there is a relationship between multiplication and division, isn't there? Here's another story for you. There were 3 horses and each one had its own stall. How many stalls were there?*
	Teacher observes students taking 3 cubes and separating them into 3 separate piles.
Nora:	*There would be 3 stalls because each of the horses is in one. We could write it as 3 ÷ 1 = 3.*
Teacher:	*Could we use multiplication to check it?*
Nora:	*Yes, 3 × 1 = 3, so 3 ÷ 1 must equal 3.*
Teacher:	*Thumbs up if you agree with Nora.*
Jim:	*That reminds me of when we were talking about dividing by 1 and it always ends up being the same number we started with.*
Teacher:	*Yes! So we have another example that supports our conjecture, don't we? Let's try another one. Jenny has no stickers. She wants to share them with 2 friends. How many stickers can each friend get?*
	Teacher notices that the students look a little confused and aren't grabbing any cubes.
Teacher:	*What do you think? How many stickers can each friend get?*
Brandon:	*Well, they can't get any because there aren't any to start with. She doesn't have any stickers.*
Teacher:	*Ah. So, how do you think we could write that as a division sentence?*
Brandon:	*0 ÷ 3 = 0*
Teacher:	*Thumbs up if you agree and thumbs down if you disagree.*
Margarite:	*I agree because you have the 0 to start and then you are sharing it between 3 people and they each get 0.*
Teacher:	*Could someone make us up a story where you have no objects?*
Stephen:	*I had no baseball cards and wanted to share them with 4 of my friends. How many can each friend get?*
Teacher:	*And how could we write that as a number sentence, Stephen?*
Stephen:	*0 ÷ 4 = 0*
Margarite:	*We could also write it as 0 = 0 ÷ 4.*
Teacher:	*Why are we allowed to do that?*
Margarite:	*Because the equal sign means "the same as." The 0 is the same as 0 ÷ 4.*
Teacher:	*Thank you for that reminder, Margarite. Is anyone noticing a pattern?*
Nora:	*Well, with the examples where there is nothing to start with, the answer is always ending up being zero.*
Teacher:	*Do you think that will always be true?*
Nora:	*I think so because you are starting with nothing, so you can share something that you don't have.*
Teacher:	*Thumbs up if you agree with Nora, and thumbs down if you disagree.*
	Teacher notices all students have their thumbs up.
Teacher:	*So, we all think that when we divide 0 by numbers, we will always have an answer of 0?*
	Students nod their heads. Brandon raises his hand.
Teacher:	*Yes, Brandon?*

Brandon: *Sometimes when I see problems like 0 ÷ 5 I get a little confused, but I think about multiplying to help me. So, when I see 0 ÷ 5 = ? I think that 5 times something equals 0 and then I know that the answer must be 0.*

Teacher: *Absolutely! Thinking about multiplication facts can really help us with the division facts!*

 SPOTLIGHT ACTIVITY

Students are often incredulous that 0 can be divided by a number. It is important for them to experience the idea. We suggest that you go back to acting out problems in the bakery. Only this time, there are no donuts to put in boxes. So, students begin to understand that the answer will always be zero (see Figure 4.14).

Story 1: The bakery had 0 cupcakes. They put all of them in 1 box. How many cupcakes did the bakery have the box?

Story 2: The bakery had 0 cupcakes. They put 0 cupcakes in each box. How many boxes did they use?

FIGURE 4.14 Bakery Boxes

Teaching zero can be so tricky. Here are a variety of activities that reinforce the concept (see Figures 4.15–4.20). Remember that students should do a lot of work with division mats so that they can see the problems in action. Students should be solving and telling word problems. They should be drawing models and using tools.

Dividing ÷0 by a Number Concrete, Pictorial, and Abstract Activities (Mixed Number Practice)
We would have students act out a few stories so that they actually understand and can explain the concept. I would have them also draw some pictures to illustrate their stories. This is a very hard concept for students to understand. They need an opportunity to talk about it and act it out and explain what it means. We would have students make a poster or write a book explaining what happens when you multiply by zero.We would also have them play different games where these facts were mixed in with other facts.

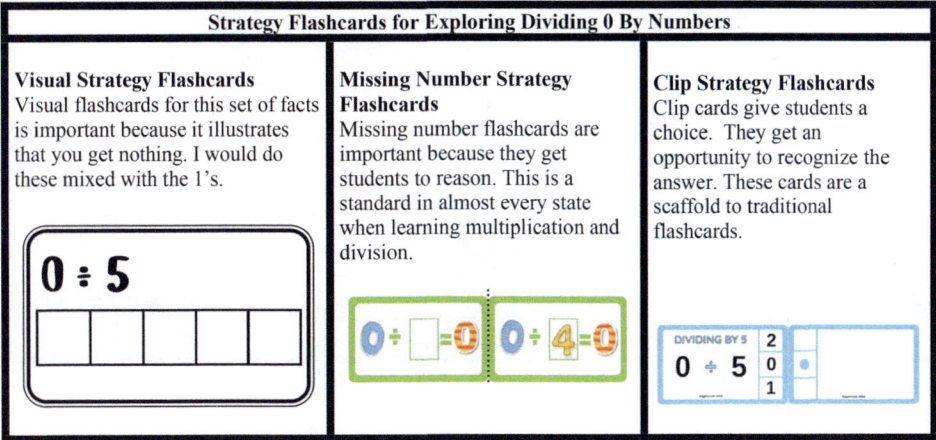

Strategy Flashcards for Exploring Dividing 0 By Numbers		
Visual Strategy Flashcards Visual flashcards for this set of facts is important because it illustrates that you get nothing. I would do these mixed with the 1's.	**Missing Number Strategy Flashcards** Missing number flashcards are important because they get students to reason. This is a standard in almost every state when learning multiplication and division.	**Clip Strategy Flashcards** Clip cards give students a choice. They get an opportunity to recognize the answer. These cards are a scaffold to traditional flashcards.

FIGURE 4.15 AND 4.16 Concrete, Pictorial, Abstract Activities and Strategy Flashcards

In every fluency module there should be a focus on word problems. Here are a few examples of the types of word problems that involve dividing 0 by numbers			
My Divide 0 By Numbers Story Problems Booklet	There were no marbles. If they were shared between two friends, how many did each friend receive? Write the set-up equation: Show your thinking with a model. Write the solution equation.	Jamal had 0 coins. If he has bags that fit 3 coins, how many bags can he fill? Write the set-up equation: Show your thinking with a model. Write the solution equation.	There were 0 rings in the seal exhibit at the zoo. If there were 4 seals, how many rings could each of them play with? Write the set-up equation: Show your thinking with a model. Write the solution equation.

FIGURE 4.17 Story Problem Booklet

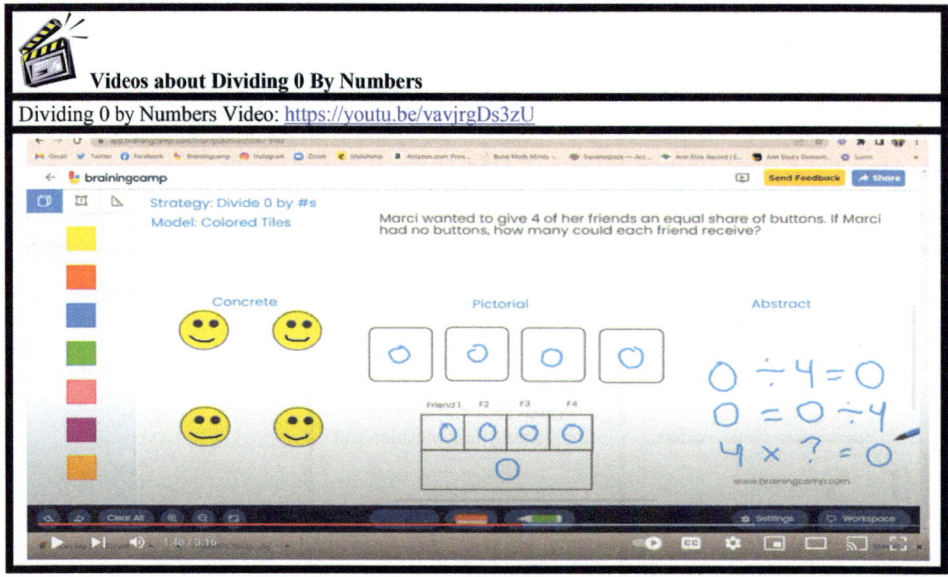

FIGURE 4.18 Resources

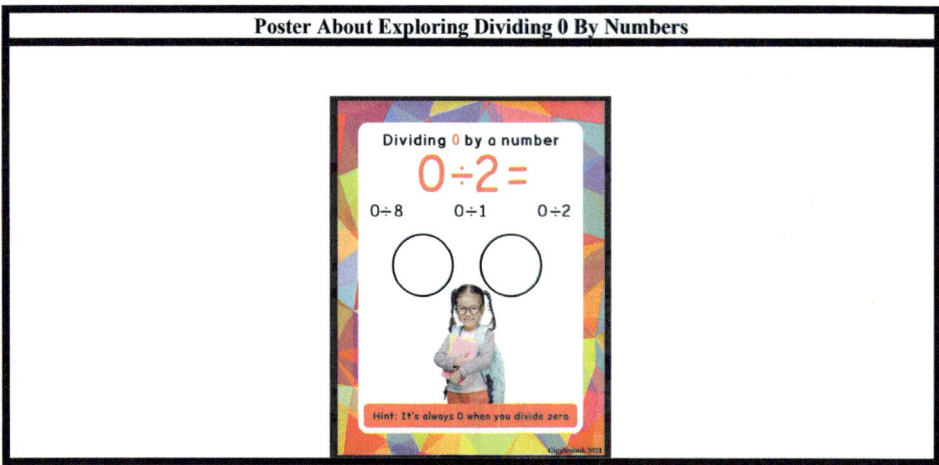

FIGURE 4.19 Posters

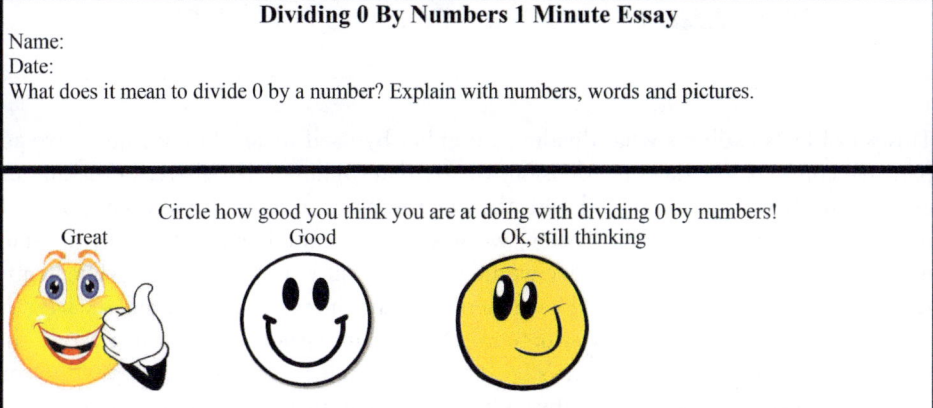

Dividing 0 By Numbers 1 Minute Essay

Name:

Date:

What does it mean to divide 0 by a number? Explain with numbers, words and pictures.

Circle how good you think you are at doing with dividing 0 by numbers!

Great   Good   Ok, still thinking

FIGURE 4.20   One-Minute Essay

EXPLORING AND LEARNING DIVIDING A NUMBER BY ITSELF FACTS

This set of facts explores what dividing a number by itself means in both quotative and partitive contexts. For partitive division, we are determining a fair share, so the amount of items we have is being shared with the same number of groups, and each group will receive 1 item. For quotative division, we would have containers that can fit the same number of items that we have, so we would be able to fill 1 container. It would be helpful to give expressions to students and have them generate the story contexts and see whether students are creating both story situations. Once again, rather than learning a rule that anything divided by itself will always be 1, we want students to have developed the conceptual understanding of why we always have a quotient of 1. This will only happen if we provide opportunities for students in a variety of contexts to explore this concept.

WHOLE-CLASS ACTIVITIES

Routines—Over, Under, the Same

In Dr. Nicki Newton's *Daily Math Thinking Routines in Action*, she shares a routine called Over, Under, the Same. During this routine, the teacher will first give students a target number such as 1. Then, the teacher will give students expressions whose value will either be less than 1, equal to 1, or over 1. At this point in the learning journey, teachers may use examples from dividing by 1, dividing 0 by anything, and dividing any number by itself. As students decide whether the answer is less than, equal to, or more than 1, they will move their arms to demonstrate their answer. If the answer is over 1, they will raise their arms to the sky, if the answer is equal to 1 their arms will be perpendicular to the ground, and if they answer is less than 1, their arms will be by their sides.

WHOLE-CLASS MINI LESSON: DIVIDING A NUMBER BY ITSELF

Teacher: *Today, we are going to work on some problems, and then we will talk about what you noticed. Here's our first problem. I have 4 pencils. I am going to share them equally among 4 kids. How many do you think each kid will get?*

Students: *1*

Teacher: *Why?*

Maria: *Because you have 4 kids and they each get 1 of your 4 pencils. There are none left.*

Teacher: *Ok, I have 5 crayons. I am going to share my 5 crayons with 5 kids. Here you go (teacher gives them out to 5 kids). How many did each kid get?*

Students: *5*

Teacher: *Why?*

Carlos: *Because you only had 5, so each person got 1.*

Teacher: Now what if I said I was going to put all these pencils in boxes that fit 5.

Kelly: You would use 1 box.

Teacher: Let's do a quick sketch of our problem. Draw 5 kids. Okay let's share the 5 pencils among them. Yes, I see everybody got 1. What would the equation be?

Luke: 5 divided by 5 is 1.

Teacher: Let's do another quick sketch of a problem. Draw 8 kids. Quick math sketch—just heads . . . okay if I had 8 pencils, show me with your math sketch of how many each kid would get.

Tariq: 1 because there are 8 pencils and 8 kids. The equation is 8 divided by 8 is 1.

Teacher: Ok, what seems to be the pattern here?

Lara: When you divide a number by itself you get 1.

Teacher: Who agrees and can give another example.

Jamal: I agree. Like if I had 3 ice creams and I had 3 kids, each kid would get 1 ice cream.

Teacher: Great, so today we have looked at what happens when you divide a number by itself. We have found that the quotient is always 1. We will continue to work on this and write stories about this situation in the next few days.

SPOTLIGHT ACTIVITY

Hula Hoop   What does it mean to divide a number by itself? Let's explore that with hula hoops. Let's start by using 1 hula hoop. If we have 1 person how many people are going to be in that hula hoop. Now let's have 2 people come up. We now have 2 hula hoops. How many people will be in each hula hoop? Ok, there will be 1 person in each hula hoop. Next let's see if we have 3 people, and 3 hula hoops, how many people will be in each hula hoop? So students begin to understand the pattern and the concept.

FIGURE 4.21   Spotlight Activity

Dividing a number by itself can be very confusing. Students will often say is it the number or 1. We have to build a capacity to reason about all problems so that students are not tethered to their memories when trying to solve problems (see Figures 4.22–4.29).

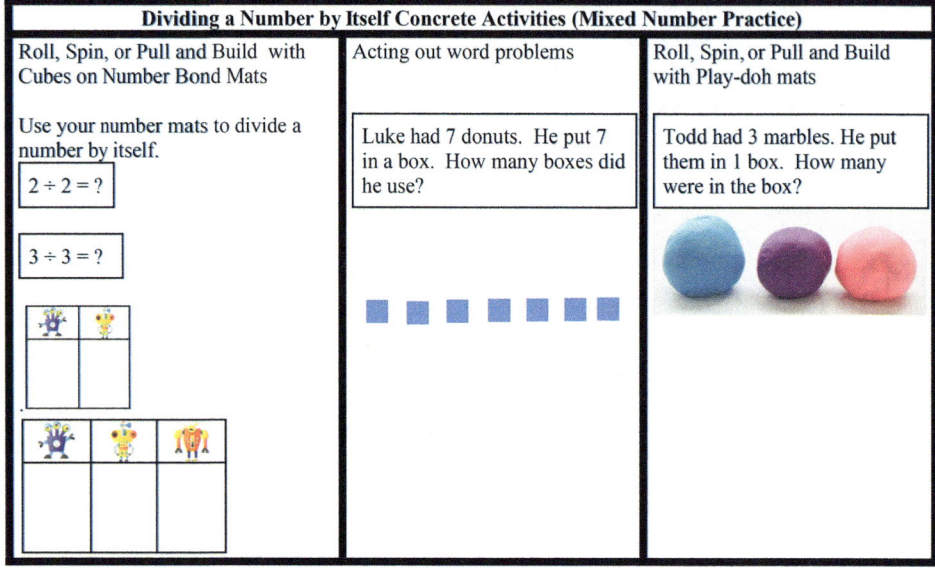

FIGURE 4.22 Concrete Activities

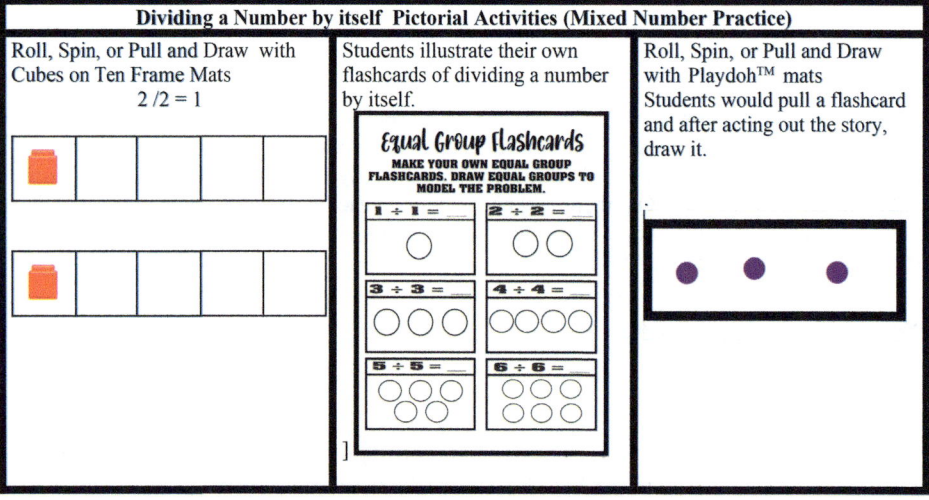

FIGURE 4.23 Pictorial Activities

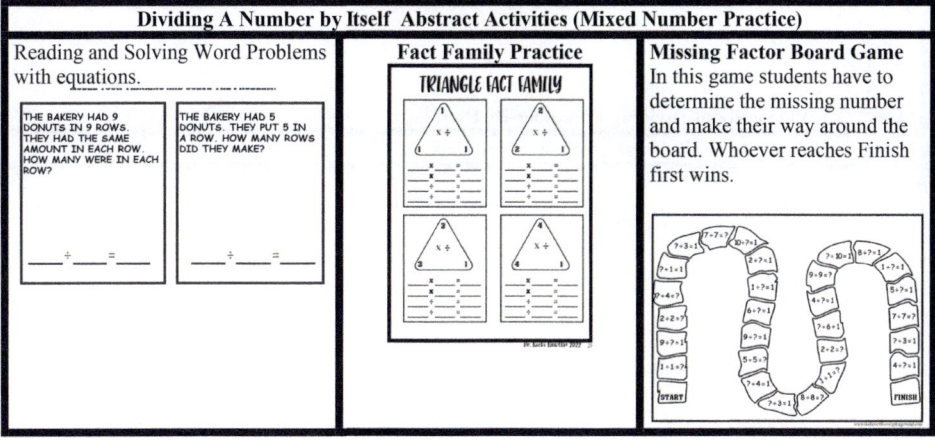

FIGURE 4.24 Abstract Activities

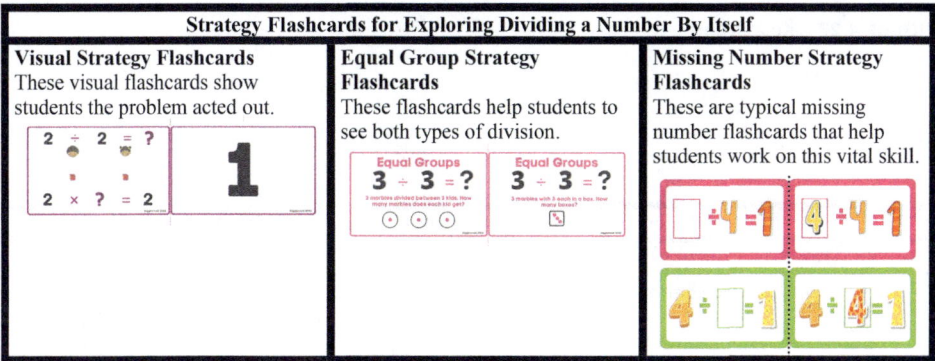

FIGURE 4.25 Strategy Flashcards

In every fluency module there should be a focus on word problems. Here are a few examples of the types of word problems that involve dividing a number by itself.			
My Dividing a Number By Itself Story Problems Booklet	Jessica wanted to give 4 ribbons to 4 of her friends. If they each received the same number of ribbons, how many did each friend receive?	A baker had 6 cookies. If she had boxes that could fit 6 cookies, how many boxes will she need to put them all in a box?	There were 9 tulip plants in a garden organized in rows that can fit 9 plants. How many rows were full of the tulip plants?
	Write the set-up equation:	Write the set-up equation:	Write the set-up equation:
	Show your thinking with a model.	Show your thinking with a model.	Show your thinking with a model.
	Write the solution equation.	Write the solution equation.	Write the solution equation.

FIGURE 4.26 Story Problem Booklet

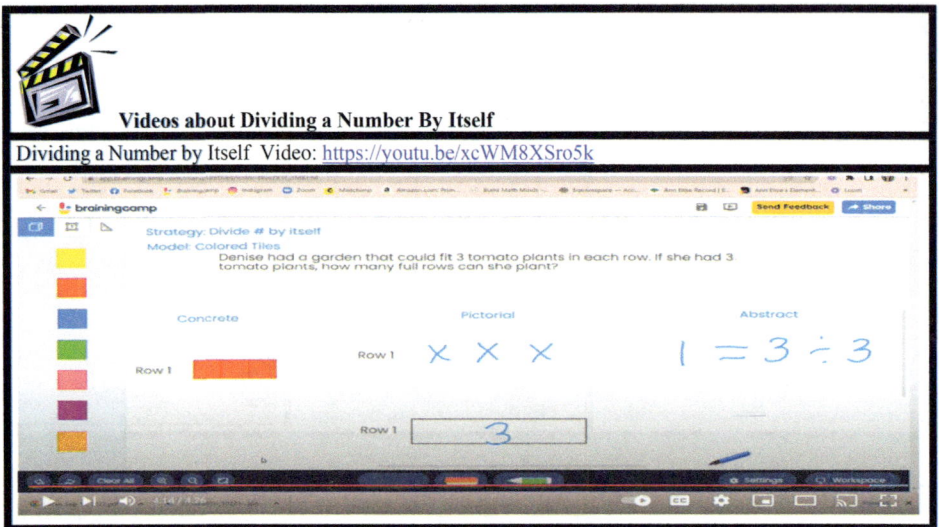

FIGURE 4.27 Resources

RESOURCES

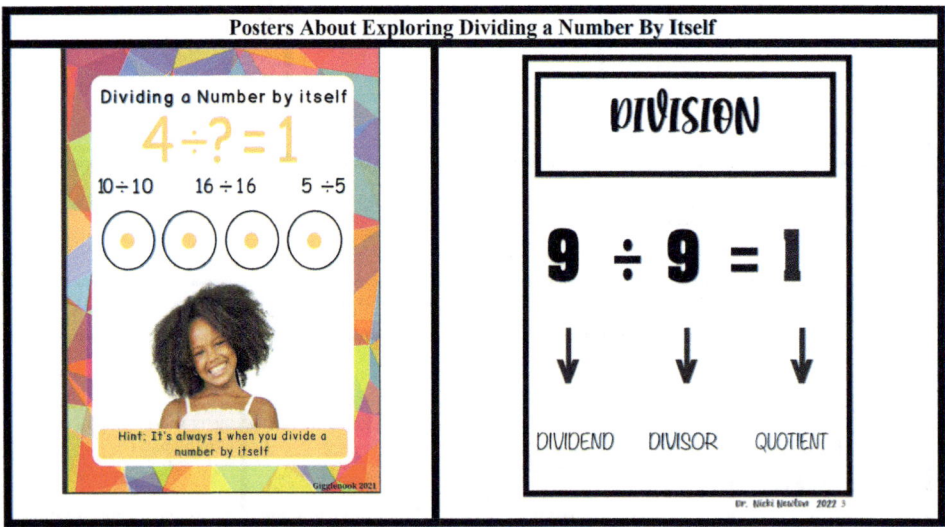

FIGURE 4.28 Posters

Dividing a Number By Itself Quiz

Name:
Date:

$2 \div 2 =$ Model with equal groups.	$6 \div 6 =$ Model with an array,	Solve. Jaden had 8 frisbees and wanted to share them between the 8 players on the team. If they all received the same amount, how many frisbees could they each get? Write the equation: Strategy: _____ Answer: _____
Solve. A farmer had stables that could fit 8 horses in each. If he had 8 horses, how many stables would be needed? Write the equation: Strategy: _____ Answer: _____	Solve. $2 \div 2 =$ _____ _____ $\div 6 = 1$ $1 = 7 \div$ _____ _____ $= 4 \div 4$	Make your own problem and solve of dividing a number by itself.

$6 \div 6 =$

Model on the number line.

What happens when we divide a number by itself? Explain with numbers, words and pictures.

Circle how good you think you are at dividing a number by itself facts!

    Great   Good    Ok, still thinking

FIGURE 4.29 Dividing by a Number Quiz

EXPLORING AND LEARNING DIVIDING A NUMBER BY 2 (IN HALF)

One of the most powerful strategies we can use when multiplying is introduced as double and halving, even though it can also be implemented for triples and thirds, quadruples and fourths, and so on. We have found, however, that we cannot assume that students are able to determine half of numbers. We can begin this thinking, though, starting with basic facts. The very same child who will tell you without hesitation that $7 + 7 = 14$, when asked $14-7$ will count back the 7. This indicates that they have not had sufficient the time to explore the inverse relationship between addition and subtraction. With multiplication and division, there is a similar inverse relationship that students need time to explore. The same factor we multiply by 2 will be the quotient when we divide its double by 2. If this thinking is developed, students are more likely to efficiently calculate products and quotients with larger numbers, decimals, and fractions.

WHOLE-CLASS ACTIVITIES

Routines—Double and Halving

The multiplication strategy of doubling and halving is a very powerful one that can be used with all sets of numbers. This routine will build capacity with halving, which is often tricky, and build the connection between the numbers. To start, the teacher simply gives a number and states whether the students need to double the number or half the number. The examples aren't limited to basic facts but can be larger numbers as well. It is powerful when we can connect the division facts to the doubling facts as well. So, a student may say, "I know that $14 \div 2 = 7$ because $7 \times 2 = 14$." The more we can connect the division facts with the related multiplication facts, the better!

WHOLE-CLASS MINI LESSON: DIVIDING A NUMBER BY 2 (IN HALF)

Teacher ensures each student has 20 linking cubes, a whiteboard, dry erase marker, and eraser.

Teacher:	*Today we are going to explore dividing amounts in half. What do you think of when I say the word "half?"*
Georgina:	*I think of sharing food with my brother. We will share the food so that each of us gets the same amount.*
Teacher:	*You mentioned something really important, Georgina. When amounts are split in half, they are divided into two equal pieces or even groups. Let's say that Marianna had 8 cookies and she was going to share them equally with Georgina. We can write this as $8 \div 2$. How many will each of them get? Let's act this out with your cubes.*
Marianna:	*We would both get 4.*

Teacher:	*How do did you figure that out?*
Marianna:	*Well, I kept giving a cube to each of the two groups and saw that there were 4 in each group when I was done.*
Teacher:	*So, we can then say that 4 is half of 8 since 2 of them make an 8. Write on your whiteboard a way we can express this situation using multiplication.*
Brandon:	*I wrote 2 × 4 = 8 because two 4's equals 8.*
Teacher:	*Yes, remember when we were working on our×2 multiplication facts and saw that they are the same as our doubles facts. So, we can say that 4 is half of 8 since two 4's make up an 8. Let's try another example. What if Brandon had 12 marbles and he was going to share them equally with Jamal? Everyone act that out using your linking cubes.*
Janet:	*I think Jamal and Brandon would each get 6 because I know that 2 × 6 = 12. I just took the pile and split it down the middle, so I had 6 in each group.*
Teacher:	*Give me our "agree" hand signal if you agree with Janet that they would each get 6.*
	Teacher waits to see how many students give the hand signal.
Teacher:	*Let's think about how we can record this situation as a division expression.*
Georgina:	*You can write 12 ÷ 2 = 6 because 12 divided into 2 equal groups is 6 in each group.*
Teacher:	*I'd love everyone to show me with your fingers how many is half of 12.*
	Teacher observes students to see if anyone is not holding up 6 fingers.
Teacher:	*Again, how do we know that 6 is half of 12?*
Marianna:	*Because it takes two 6's to make 12.*
Teacher:	*Would you all please write the division sentence and the multiplication sentence for this situation?*
Brandon:	*I wrote 12 ÷ 2 = 6 and 2 × 6 = 12.*
Teacher:	*Give the "agree" hand signal if you agree.*
Teacher:	*I'd like you to think of a number up to 20 and then figure out how many is half of it by proving it with your linking cubes.*

Teacher observes students to see which students successfully hold up 8 fingers to determine any students that may need more practice in a small-group setting the next day.

SPOTLIGHT ACTIVITY

Read the book *The Great Divide* by Dayle Ann Dodds to the students (www.youtube. com/watch?v=7ZSlOqD5h6U) and discuss what is happening when we are determining half of a number by dividing by 2. Include in your discussion the connection to multiplication and thinking that 2 times something would be the number being halved. Then, provide a variety of numbers on the board that can be halved at least a couple of times and have students choose a number and continue to half the resulting numbers, mirroring what happened in the book until they get the lowest whole number (and further for those students who can think through halving fractions).

Dividing by 2 is a relatively easy concept if students understand doubling and halving. Students should practice sharing lots of different amounts between two people. Here, we have listed several activities to help you throughout the year (Figures 4.30–4.37).

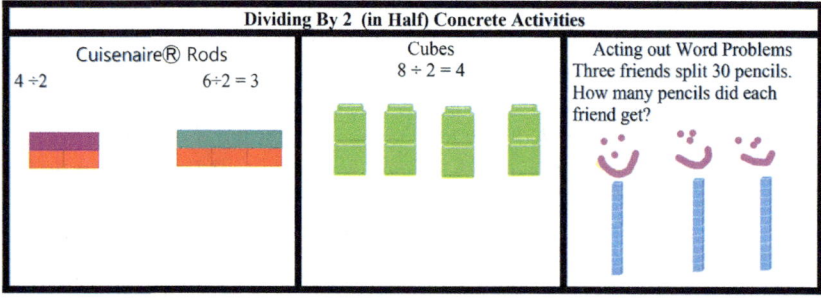

FIGURE 4.30   Concrete Activities

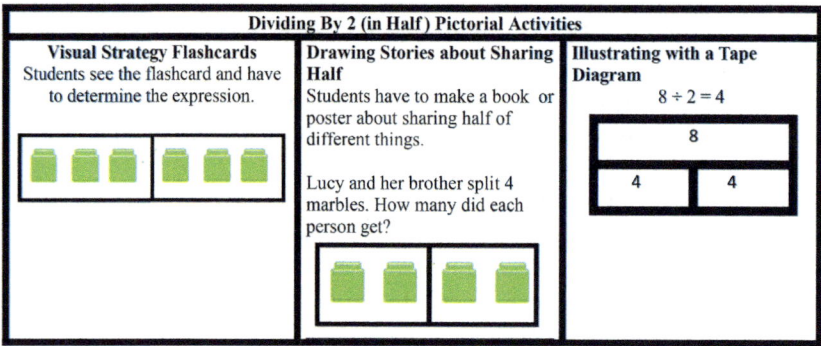

FIGURE 4.31   Pictorial Activities

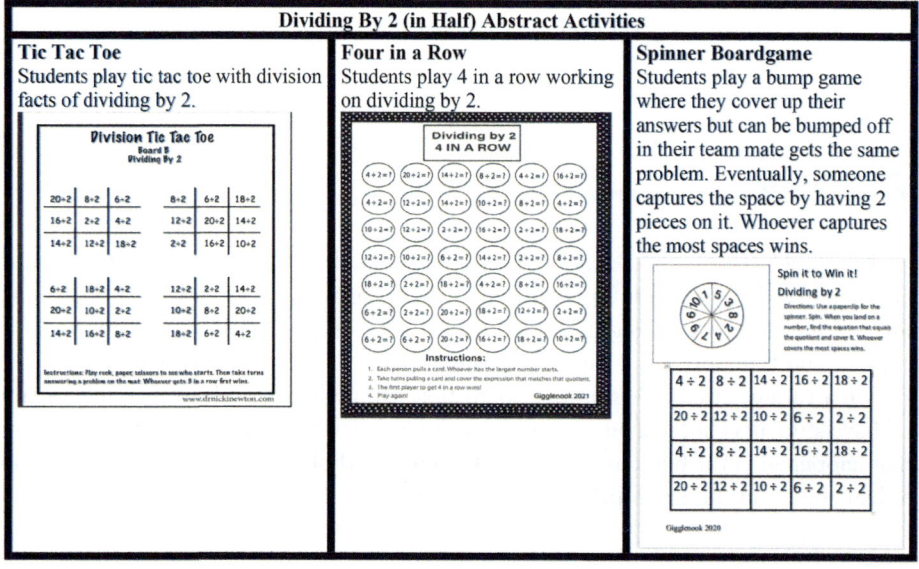

FIGURE 4.32   Abstract Activities

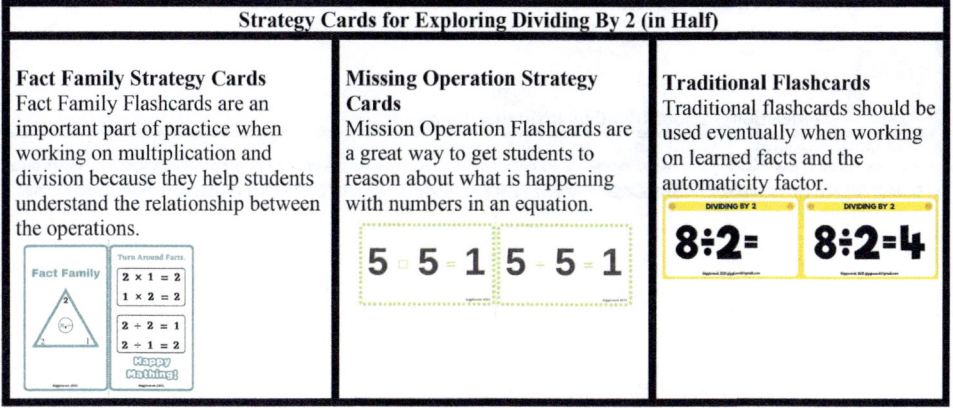

Strategy Cards for Exploring Dividing By 2 (in Half)		
Fact Family Strategy Cards Fact Family Flashcards are an important part of practice when working on multiplication and division because they help students understand the relationship between the operations.	**Missing Operation Strategy Cards** Mission Operation Flashcards are a great way to get students to reason about what is happening with numbers in an equation.	**Traditional Flashcards** Traditional flashcards should be used eventually when working on learned facts and the automaticity factor.

FIGURE 4.33 Strategy Flashcards

In every fluency module there should be a focus on word problems. Here are a few examples of the types of word problems that involve dividing a number by 2 (in half)			
My Dividing a Number By 2 (in Half) Story Problem Booklet	Gabrielle had 14 towels and she wanted to put them into 2 equal piles. How many will be in each pile? Write the set-up equation: Show your thinking with a model. Write the solution equation.	There were 16 cars in a parking lot. If the cars are parked equally between two rows, how many cars are in each row? Write the set-up equation: Show your thinking with a model. Write the solution equation.	Jillian had twice as much money as her sister Janelle. If Jillian has $18, how much money does Janelle have? Write the set-up equation: Show your thinking with a model. Write the solution equation.

FIGURE 4.34 Story Problem Booklet

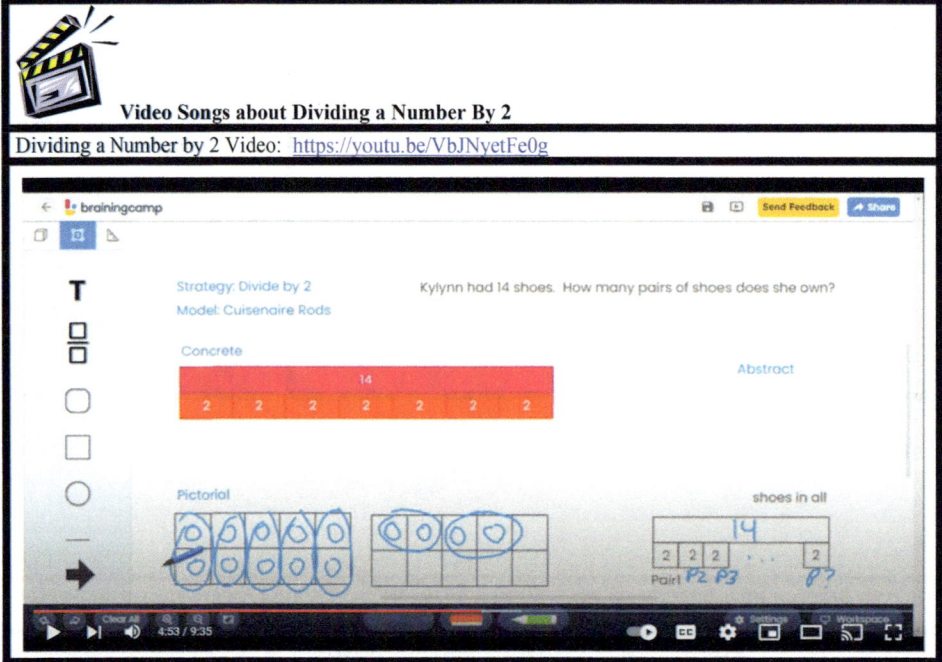

FIGURE 4.35 Resources

RESOURCES

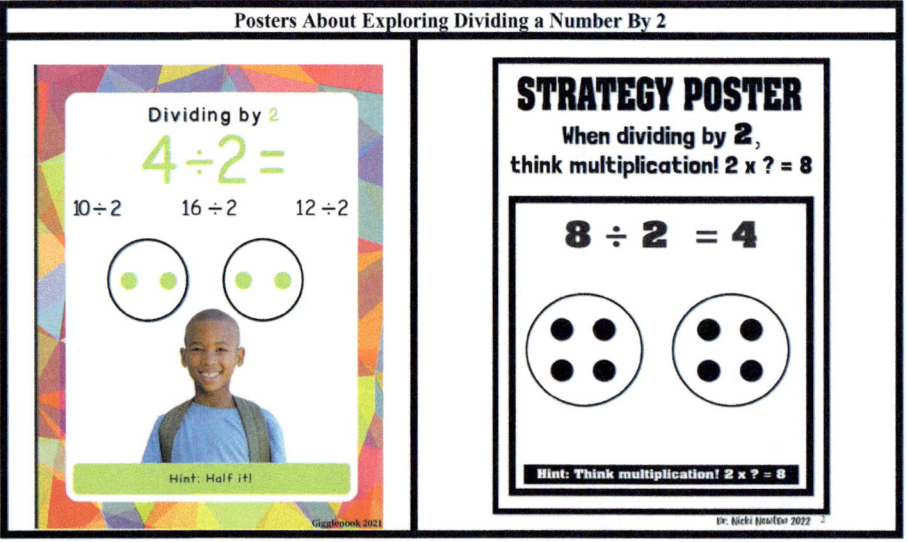

FIGURE 4.36 Posters

Dividing a Number By 2 Quiz

Name:
Date:

$12 \div 2 =$ Model with equal groups.	$8 \div 2 =$ Model with an array.	Solve. Zaki had 16 toy cars. He wanted to store them in a container that fits 2 toy cars to a section. How many sections will be full if he puts all his cars in the container? Write the equation: Strategy: _____ Answer: _____
Solve. Janice had half as many barrettes as Kylynn. If Kylynn has 14 barrettes, how many does Janice have? Write the equation: Strategy: _____ Answer: _____	Solve. $12 \div$ _____ $= 6$ _____ $\div 2 = 9$ $3 = 6 \div$ _____ _____ $= 10 \div 2$	Make up your own problem dividing by 2 and solve.

$8 \div 2 =$

Model on the number line.

What happens when we divide a number in half? Explain with numbers, words and pictures.

Circle how good you think you are at dividing by 2 facts!

Great	Good	Ok, still thinking

FIGURE 4.37 Dividing by a Number Quiz

EXPLORING AND LEARNING DIVIDING A NUMBER BY ITS HALF FACTS

This set of facts is also related to the doubles facts, only this time, the divisor is the amount that is half of the dividend. If we think about how students have typically progressed with facts in sequential order, a fact like $14 \div 7$ would be included within the divide by 7 facts. During many interviews with students using Math Running Records, the expressions with a 7 are frequently mentioned as some of the trickiest of the expressions no matter what the operation. Yet, when we take the time to work on half facts as their own set of facts and connect $14 \div 7$ to $7 \times 2 = 14$, that fact that once looked tricky will then look easy because it becomes familiar. This is a critical piece of building positive math identities. When asked a fact, we want students to think about what they *do* know to help them solve the problems that they *don't* know. They will develop the disposition that while they may not know the fact, they do have the thinking strategies that will help them figure it out.

WHOLE-CLASS ACTIVITIES

Routines — Disappearing Dan

Daily Math Thinking Routines in Action (Newton, 2018) features a routine called Disappearing Dan. In this routine, the teacher draws a person (see Figure 4.38) on the board and then puts numbers on various sections of the body. The teacher will then ask a student a question based on the information on the board. Once the student correctly answers the question, that part of the body is erased. As questions are answered, Dan is disappearing. Since the teacher is customizing the question to the student who selects the number, this routine provides an opportunity to differentiate and meet students where they are.

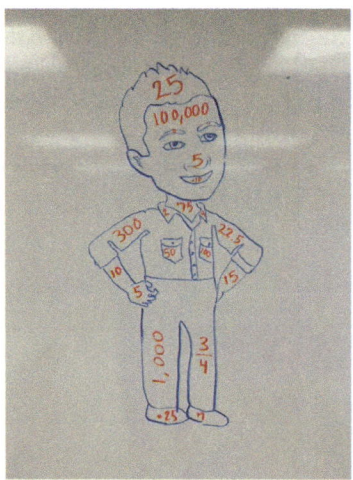

FIGURE 4.38 Disappearing Dan

WHOLE-CLASS MINI LESSON: DIVIDING BY A NUMBER BY ITS HALF

Each student should be provided with a student kit of Cuisenaire® rods, a whiteboard, marker, and eraser. Special note: we are including images of Cuisenaire® rods with labels on them only for the readers' benefit in case they are unfamiliar with the rods (see Figures 4.38–4.40). We do not use labeled Cuisenaire® rods with students.

Teacher:	*If anyone wants to build a staircase of your Cuisenaire rods, go right ahead. Remember you can also refer to our poster of the staircase on the wall as well. If you don't remember how long a rod is, think about any of the rods you do know and use that to help you. Let's begin with this story situation: There were some containers of pencils that fit 5 pencils. How many boxes would be needed if there were 10 pencils? Let's prove it with our Cuisenaire rods.*
Teacher:	*Let's grab a 10 rod and lay that down horizontally. If we had one box of pencils, how many pencils would that be?*
Susan:	*You said each box fits 5 pencils, so that's 5 pencils.*
Teacher:	*That's right. Let's see how many groups of 5 there are in 10 and show that to me using your Cuisenaire rods.*

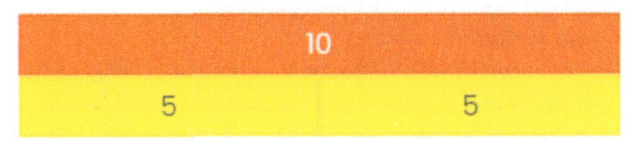

www.brainingcamp.com

FIGURE 4.38A Model 1

Joey:	*It took 2 of the groups of 5, so then they would need 2 boxes to fit all the pencils.*
Teacher:	*Super! I'm going to record this on the board for us. The original problem was 10 ÷ 5 = ? and we just figured out that 2 × 5 = 10, so now we can also record that 10 ÷ 5 = 2.*
	Let's try another situation. A farmer has 12 sheep. She places them in pens that fit 6 sheep each. How many pens will be full? What division equation can we write that would represent this situation?
David:	*12 ÷ 6 = ?*
Teacher:	*I'd like you to build with your Cuisenaire rods.*
	Students explore using the Cuisenaire rods while teacher observes.

www.brainingcamp.com

FIGURE 4.39   Model 2

Jessica: Two groups of 6 fit in the 12, so she would have 2 full pens.

Teacher: Let's record that 12 ÷ 6 = 2. What multiplication problem could we also write for this situation?

Joey: I knew it was going to be 2 times because when we had 12 ÷ 6 = ? I was thinking that something times 6 is 12. I just thought of my doubles facts so 2 × 6 = 12.

Teacher: If you agree with Joey, show me your "agree" hand signal.

Teacher: I have another story for you. There were 18 muffins at a bakery and the baker had boxes that fit 9 muffins. How many boxes will he need to put all the muffins in a box? I'd like you all to write a division equation on your whiteboard that represents this situation and then check with a partner.

   Students work together to write the equation.

Teacher: Who wants to share?

Susan: We thought about 18 ÷ 9 = ? because we are trying to figure out how many 9's are in 18. We know that 2 × 9 = 18 so we know there will be 2 boxes that are full.

Teacher: Let's build this with Cuisenaire rods to prove that there are two 9's in 18.

www.brainingcamp.com

FIGURE 4.40   Model 3

Joey: I'm noticing a pattern with the stories we have done today. The answer for all of them is 2.

Teacher: I'm glad you noticed that, Joey. I wanted us to focus on these division facts that all answer 2. How do you think I chose my numbers for the examples today?

Jessica: The numbers we were dividing by were half of the larger amount.

Teacher: Exactly. I was dividing the numbers by the amount that is half of it. Work together with your partners to build me more examples of numbers being divided by its half and prove to me that the answer will always be 2.

   As students are working, the teacher can record on the board all the possibilities that the students built using both the division equations and the multiplication ones.

SPOTLIGHT ACTIVITY

Herding Game

We can again play the herding game, but this time we will play it using division and exploring both dividing the herd in half and dividing the herd by its half (Meg from Teacher Studio discusses the possibilities of this game in depth). Begin by selecting a number of children, such as 8 students, to stand up in the front of the room. Then, explain that a shepherd was herding her sheep and arranged her 8 sheep into 2 equal-sized groups. The students then will arrange themselves until they are in the 2 equal groups. Record this using pictures and an equation on the board. Then, have them move back into one group, and then share that the shepherd organized the sheep into 4 equal groups and have the students arrange themselves accordingly. Continue this with other numbers and have the students look for the patterns emerging between the amount that is half the original size and the quotient when the same number is divided by its half. There are many activities that you can do throughout the year to get your students to understand these concepts (see Figures 4.41–4.48).

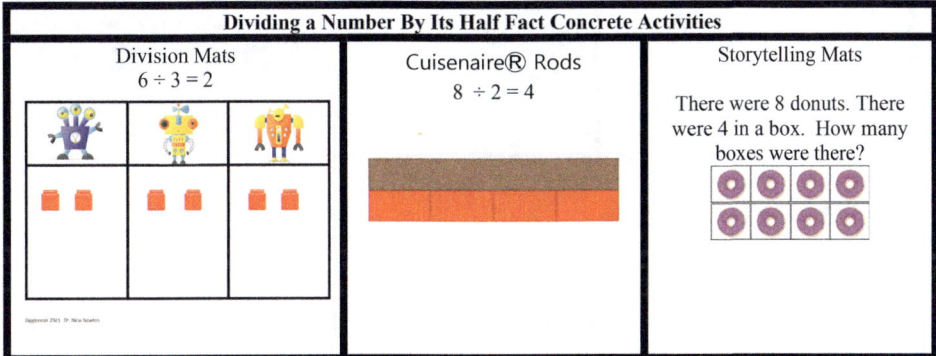

FIGURE 4.41 Concrete Activities

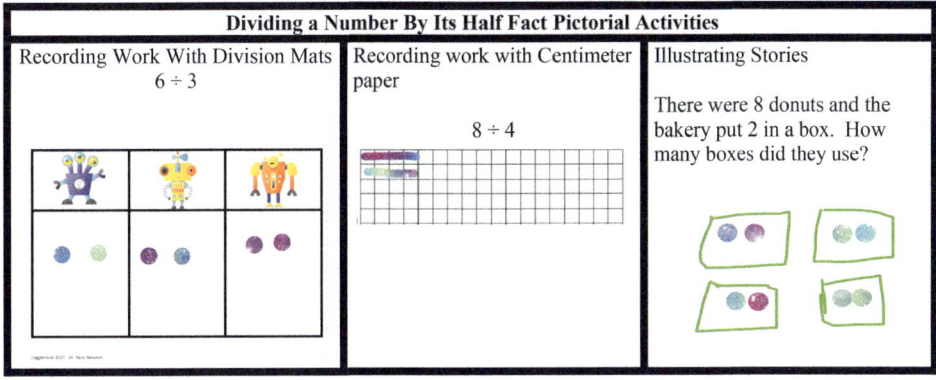

FIGURE 4.42 Pictorial Activities

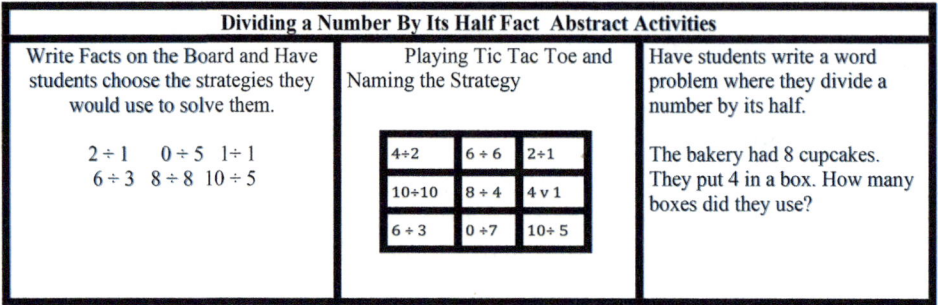

Dividing a Number By Its Half Fact Abstract Activities		
Write Facts on the Board and Have students choose the strategies they would use to solve them. 2 ÷ 1 0 ÷ 5 1 ÷ 1 6 ÷ 3 8 ÷ 8 10 ÷ 5	Playing Tic Tac Toe and Naming the Strategy	Have students write a word problem where they divide a number by its half. The bakery had 8 cupcakes. They put 4 in a box. How many boxes did they use?

FIGURE 4.43 Abstract Activities

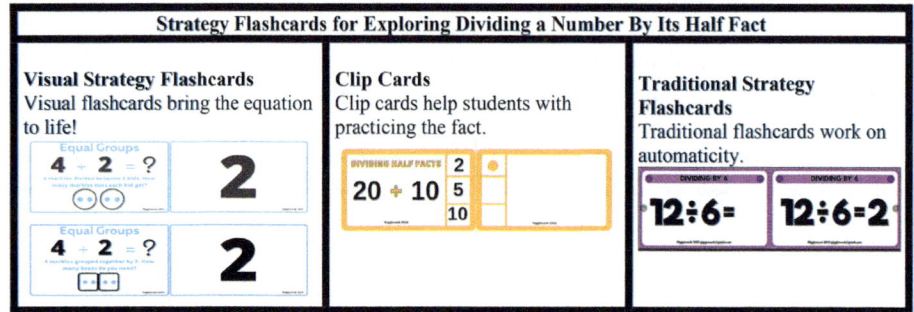

Strategy Flashcards for Exploring Dividing a Number By Its Half Fact		
Visual Strategy Flashcards Visual flashcards bring the equation to life!	**Clip Cards** Clip cards help students with practicing the fact.	**Traditional Strategy Flashcards** Traditional flashcards work on automaticity.

FIGURE 4.44 Strategy Flashcards

In every fluency module there should be a focus on word problems. Here are a few examples of the types of word problems that involve dividing a number by its half facts.			
My Dividing a Number By Its Half Fact Story Problems Booklet	Gloria had some containers of water bottles. If the containers fit 8 water bottles, and there were 16 bottles in all, how many containers did she have? Write the set-up equation: Show your thinking with a model. Write the solution equation.	There were 18 chocolate chip cookies and 9 sugar cookies on a platter. How many times more chocolate chip cookies were there than sugar cookies? Write the set-up equation: Show your thinking with a model. Write the solution equation.	Danny brought 12 dog treats to the dog park. If there were 6 dogs at the park, how many can Danny give each dog if they were shared equally? Write the set-up equation: Show your thinking with a model. Write the solution equation.

FIGURE 4.45 Story Problem Booklet

RESOURCES

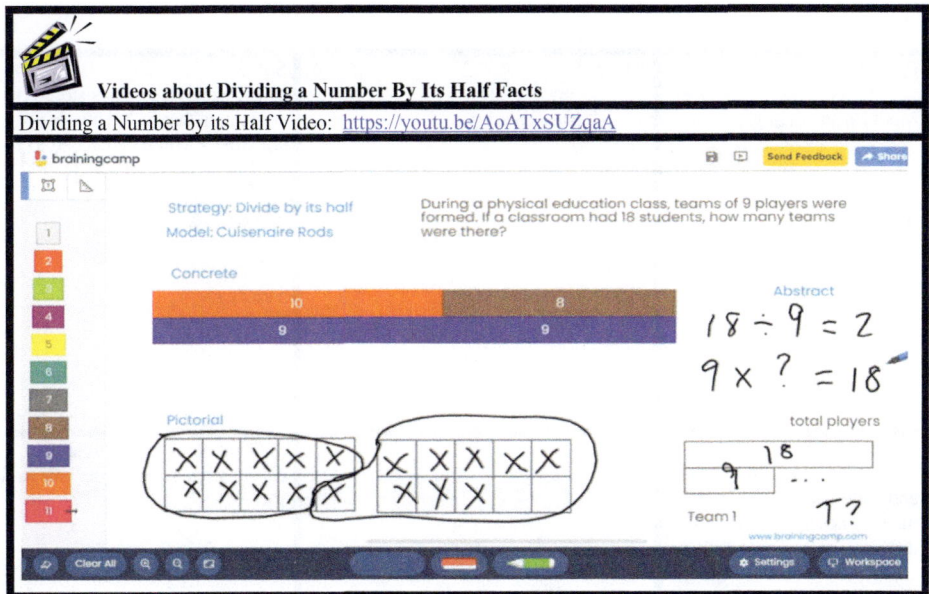

FIGURE 4.46 Resources

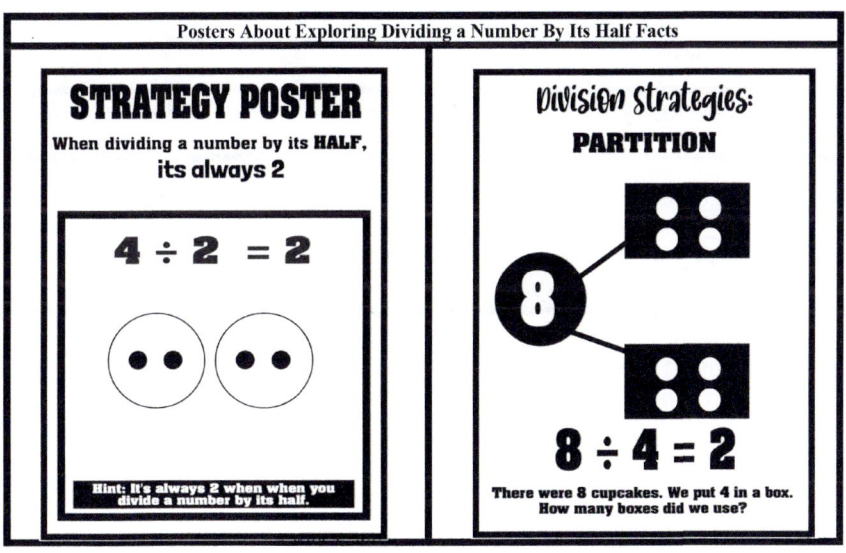

FIGURE 4.47 Posters

Dividing a Number By Its Half Fact Quiz

Name:
Date:

6 ÷ 3 = Model with equal groups.	10 ÷ 5 = Model using an array.	Solve. There were 16 earrings stored in containers that fit 8. How many containers would be needed to fit all the earrings in a container? Write the equation: Strategy: _____ Answer: _____
Solve. Toby collected 18 sea shells and Kate collected 9. How many times more sea shells did Toby collect than Kate? Write the equation: Strategy: _____ Answer: _____	Solve. 12 ÷ _____ = 2 _____ ÷ 8 = 2 _____ = 16 ÷ 8 2 = _____ ÷ 9	Make up your own problem dividing a number by its half and solve.

10 ÷ 5=

Model on the number line.

What happens when we divide a number by its half? Explain with numbers, words and pictures.

Circle how good you think you are at doing dividing by its half facts!
Great   Good   Ok, still thinking

FIGURE 4.48 Dividing a Number by its Half Quiz

EXPLORING AND LEARNING ABOUT THINKING MULTIPLICATION TO SOLVE DIVISION FACTS

As we explore the rest of the division facts, Dr. Nicki Newton suggests a progression of ÷5, ÷10, ÷3, ÷6, ÷4, ÷8, ÷9 (Newton, 2016). Once again, we are building on what students already know to help them develop mastery of more advanced facts. Exploring those connections and relationships is a crucial foundation. As mentioned previously, research has shown that the thinking that students do when asked division facts relies heavily on them knowing their multiplication facts (Kouba, 1989; Mulligan & Mitchelmore, 1997; National Research Council, Mathematics Learning Study Committee, 2001). If students have mastered their multiplication facts, we want to provide experiences for them to develop their connection between the related multiplication facts. When asked a division fact such as 48 ÷ 8, they can think "8 times something is 48." If they are able to retrieve the missing fact from long-term memory, they will be all set. What we are interested in this section of the chapter, though, is the thinking of the students who aren't able to retrieve that fact from memory. If students are using immature counting strategies, we want to provide experiences, preferably through games and activities, that support their transition to multiplicative reasoning.

What exactly does multiplicative reasoning look like in terms of basic facts? At the heart of many strategies for multiplication facts is the distributive property of multiplication. We can decompose one factor and distribute it to the other factor. Once we combine those two partial products, we will have derived the original product. For division, we can activate this same thinking in pursuing partial quotients. For example, if a student is asked that same 48 ÷ 8, rather than counting or skip counting, they can think, "What fact of 8 do I know for sure?" Many students have mastered their 5 facts, so they typically know that 5 × 8 = 40. So, rather than skip counting by 8's five times, we know that 5 groups of 8 is 40. Students can then think of how much more is in the 48 beyond that 40? Well, there is one more 8. So, 48 ÷ 8 must be 6.

Here is a series of visuals of an area model that can help us make this thinking visible (see Figure 4.49). In the first example, we can discuss how the area model can represent our thinking for division. If every multiplication expression can be represented using a rectangle, with division we are finding the largest rectangle we can make with an area matching the dividend. The vertical distance (width) is the divisor, and once we have used as much of the dividend as we can, the quotient will be the horizontal distance (length) of the rectangle.

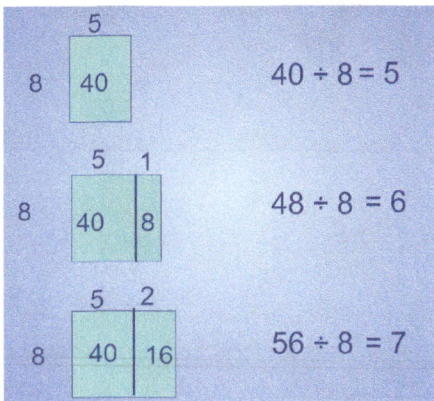

FIGURE 4.49 Model 4

WHOLE-CLASS ACTIVITIES

Routine—Steve Wyborney's Splat!

Steve Wyborney has an entire set of Splat! slide decks (https://stevewyborney. com/2017/02/splat/) ready to use for free from kindergarten to high school. The set that will help with the development of thinking multiplication to solve division is the multiple Splat slides. The Splat! routine develops over a series of slides using the animation feature within PowerPoint so that it guides the students through the discussion prompts. At first, they will see all the blue dots and are asked to determine how many there are. Then, some of the dots are covered in Splat! images. In these versions, there are multiple splats that are all the same color, which means they are each hiding the same number of dots. In other versions, there are different-colored splats, which means they are hiding different amounts of dots. In this example, we can see from the number in the upper right corner that there are 19 blue dots in the image and all but 7 are hidden. We can determine, then, that 12 dots are hidden beneath the 4 splats. At this point, students may think $12 \div 4 = ?$ or $4 \times ?$ = 12. By having students share their thinking, all students can be exposed to both ways of thinking and develop the understanding of the inverse relationship between multiplication and division. As a side note, think of the power of this image in terms of the algebra notation. It represents $4x + 7 = 19$. Crazy powerful!

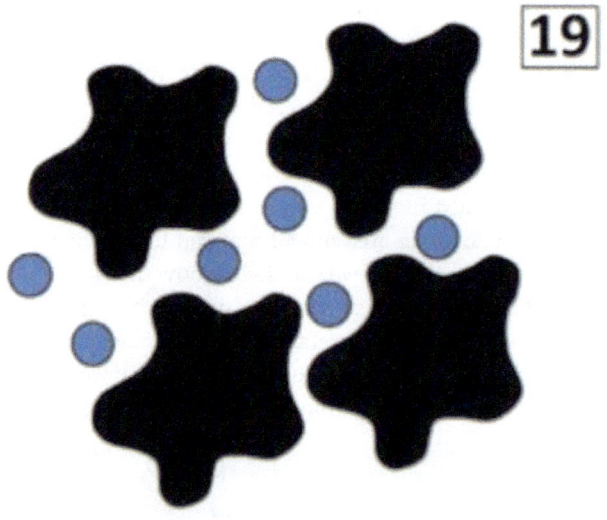

https://stevewyborney.com/2017/02/splat/

WHOLE-CLASS MINI LESSON: THINKING MULTIPLICATION WHEN DIVIDING

(Teacher ensures each student has grid paper in a plastic sheet protector, a dry erase marker, and an eraser.)

Teacher: *Today, we are going to talk about how we can think multiplication to solve division situations. Just like we do in other topics in math, we can use what you know to determine what you don't yet know. Let's begin with 56 ÷ 7. Does this look tricky to any of you?*

        Several students raise their hands.

Teacher: *I know from interviewing you with Math Running Records that 7's look pretty scary no matter what the operation we are doing. But remember when we explored that ×7 can be thought of as ×5 + ×2 and then those problems that once looked scary began to look easier to you? That same thing can happen with thinking multiplication when we divide even if we are dealing with a 7. Can anyone express this situation for us using multiplication rather than division?*

Onyx: *7 times something equals 56.*

        Teacher writes 7 × ? = 56 on the board and records as an area model as well.

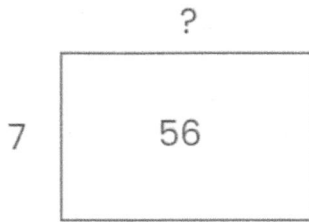

Teacher: *So, if we know our multiplication facts, we would be able to determine the answer and be done. But let's think about a student who doesn't know the answer. What could they do to figure this out?*

Christine: *They could skip count by 7's and keep track of how many sevens there are in the 56.*

Teacher: *Absolutely! Any other ideas?*

Imani: *They also could also subtract 7 from 56 over and over again to see how many 7's there are in 56.*

Teacher: *You sure can. I'm wondering if we could still figure out how many 7's there are in 56 using larger groups of 7 that you know for sure so we can save ourselves some time. Can anyone tell me a multiplication math fact of 7 that you know for sure?*

Onyx: *I know that 7 × 5 = 35.*

Teacher: *Super! So, let's record what we know so far using an area model.*

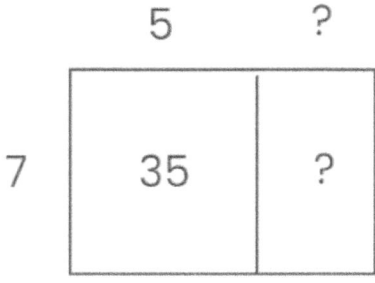

Teacher: *If we know that the total area needs to be 56 and we already have 35, how much of an area do we need to work with?*

Christine: Well, if it were 20 more than the 35, then that would be 55 so it is just one more than that. 21.

Onyx: And I know that $7 \times 3 = 21$, so we can fit 3 more groups of 7 in the 56.

Teacher: Let's take a look at that with our area model.

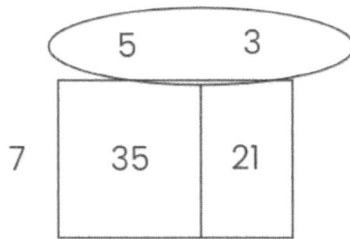

Teacher: Our area is a total of 56 and that was made up of 5 groups of 7 and then 3 more groups of 7 for a total of 8 groups of 7. So, can someone give me the completed division equation?

Imani: $56 \div 7 = 8$.

Teacher: We can also represent that sentence in this way:

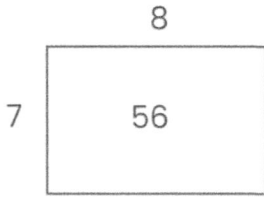

Teacher: I want to show you another way we can represent this situation. What do you notice?

$$7 \overline{\smash{)}\,56} ^{8}$$

Christine: The symbol looks just like the corner of the rectangle.

Teacher: Yes! That's where that symbol originates from. So anytime, you are asked a division problem, you can always think multiplication to figure out the answer. Always build from what you know to help you. Today we built from Onyx's knowing that $7 \times 5 = 35$, but we could have started with any fact that you know and work from there.

SPOTLIGHT ACTIVITY

The Doorbell Rang by Pat Hutchins

www.youtube.com/watch?v=A-tqjCPnHLg

Read the book *The Doorbell Rang* to your students and either have the students act out what happens each time the doorbell rings using paper cookies or small objects such as linking cubes or one-inch square tiles. Record the results on the board using a division equation and then discuss and write it as a multiplication equation to emphasize the relationship between the two as well as how we can use thinking multiplication to solve division situations. You can extend the story to imagine that there is a larger group of cookies and give scenarios with various amounts of people sharing the cookies, but predict what the answer will be before acting out the story and thinking multiplication. Once you have a prediction, then students can act out the situation and see if they were correct. This is one of many great anchor texts that you could use throughout your exploration. We have many ideas listed here to help you on your journey (see Figures 4.50–4.56).

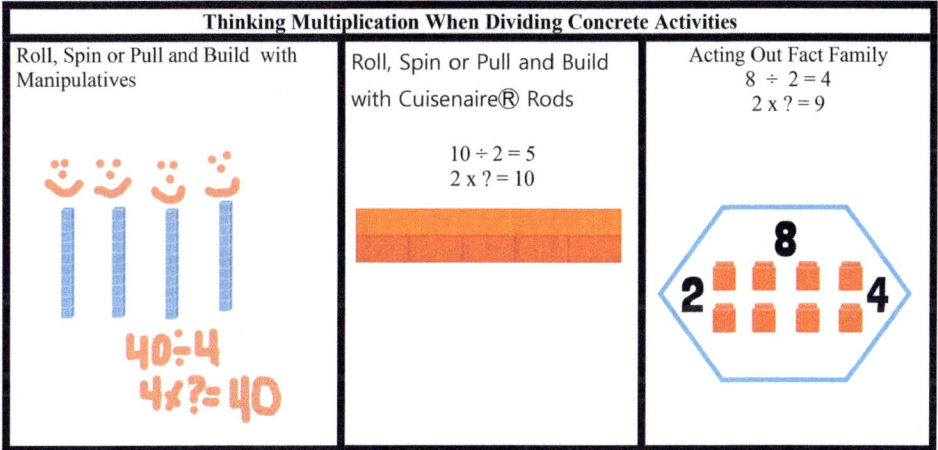

FIGURE 4.50 Concrete Activities

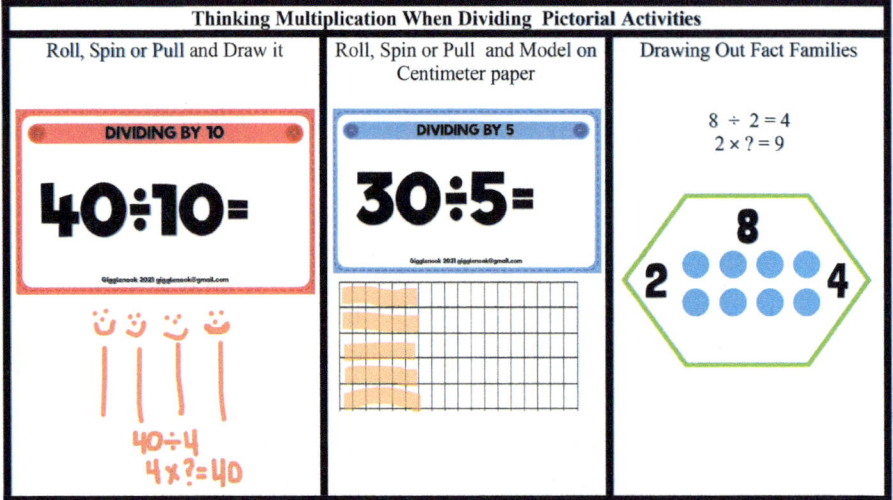

FIGURE 4.51   Pictorial Activities

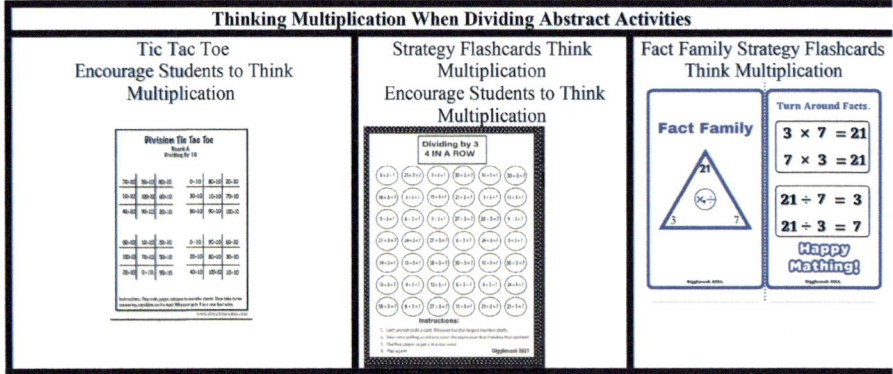

FIGURE 4.52   Abstract Activities

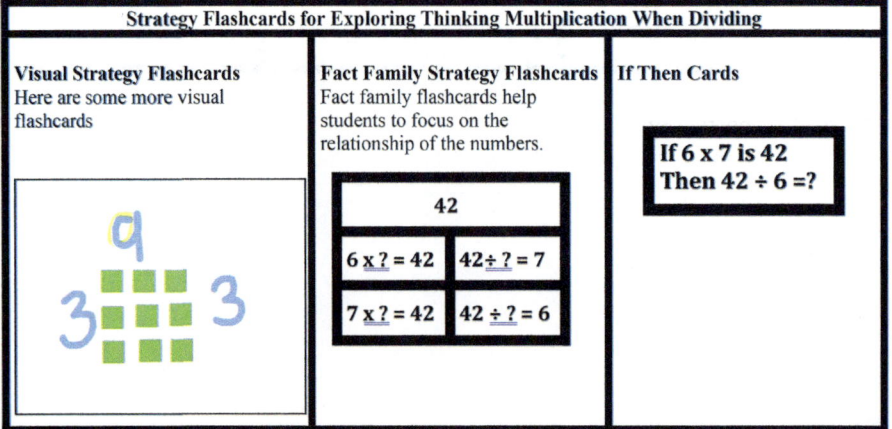

FIGURE 4.53   Strategy Flashcards

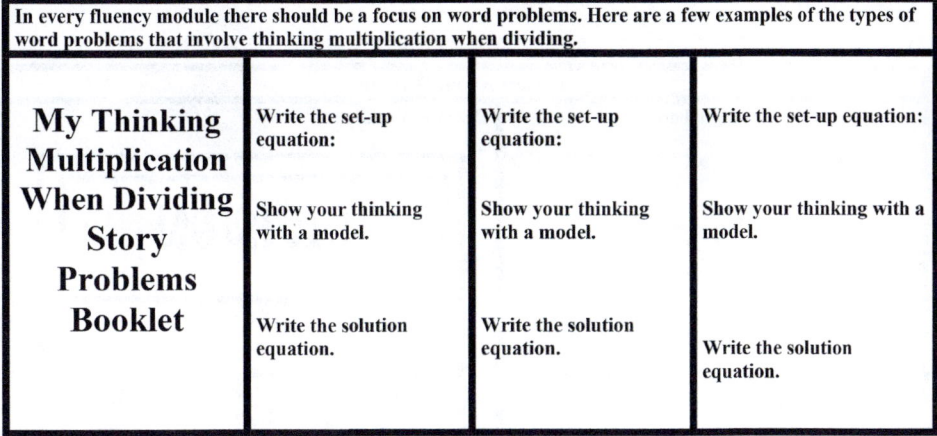

In every fluency module there should be a focus on word problems. Here are a few examples of the types of word problems that involve thinking multiplication when dividing.			
My Thinking Multiplication When Dividing Story Problems Booklet	Write the set-up equation: Show your thinking with a model. Write the solution equation.	Write the set-up equation: Show your thinking with a model. Write the solution equation.	Write the set-up equation: Show your thinking with a model. Write the solution equation.

FIGURE 4.54   Story Problem Booklet

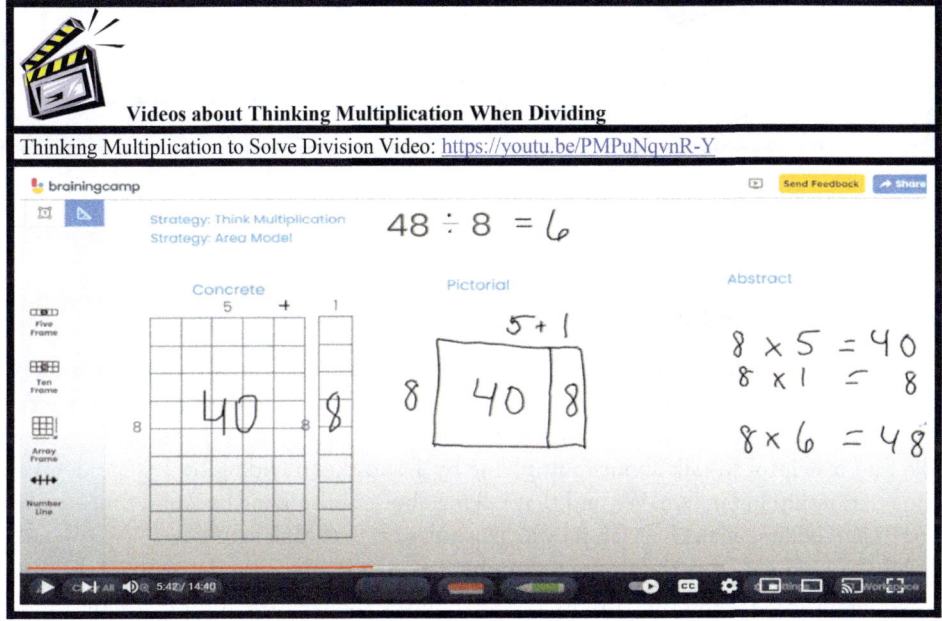

FIGURE 4.55   Resources

RESOURCES

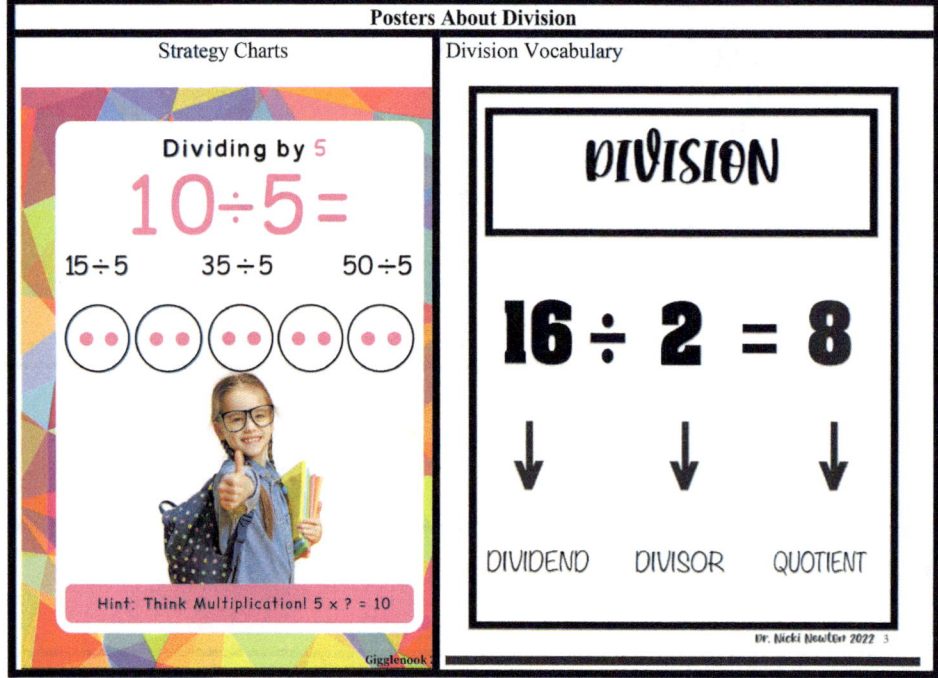

FIGURE 4.56   Posters

Digging Deeper Into Division

Once students understand the basic concepts of division, the focus is to think multiplication. Thus, it is very important that students understand multiplication before we rush them to division. We should also always try to make connections between the operations. We find it helpful to talk about multiplying by 1 and then dividing by 1 so students see the relationship in tandem. We find that talking about multiplying by 2 and then showing arrays and talking about fact families helps students to see this relationship. We are a fan of all things visual (if you didn't notice!). So, when working with fact families, we use a visual flashcard first and then eventually work in the triangle. We got this wonderful idea from Math Cats and have been making these with students ever since. We make hexagon flashcards because that is how we first saw them, and they are available for free (already with the numbers set up) on the website (www.mathcats.com/explore/factfamilies/print-multcards.html). See Figure 4.57 for examples of our cards.

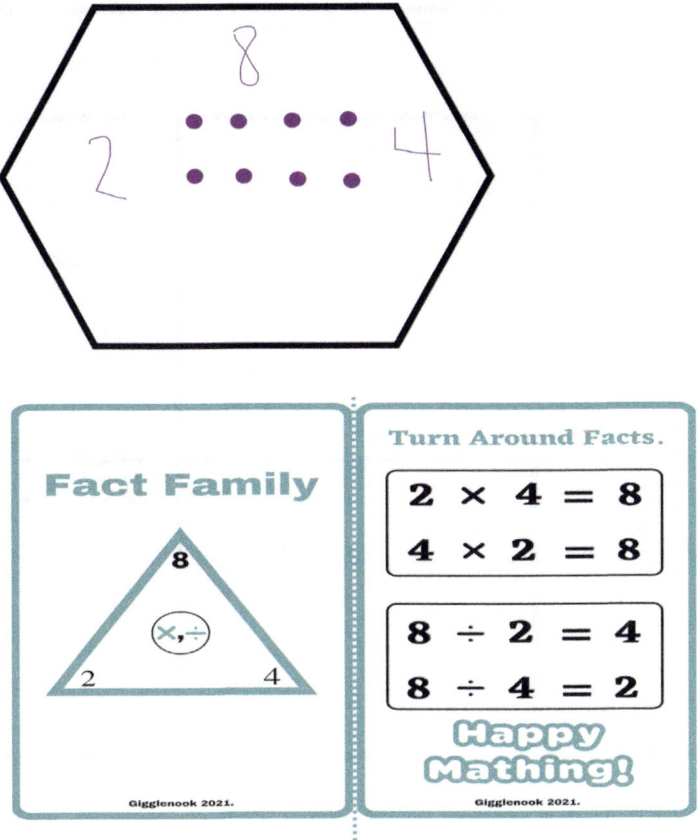

FIGURE 4.57 Flashcards

Okay, we know it is just an array. But if you are trying to teach fact families, this is the way to go! We can see the multiplication facts AND the division facts, conceptually!

   2 × 4 and 4 × 2 . . . Look from side to side or up and down (Voila! There's the multiplication facts).
   8 ÷ 2 . . . Split it horizontally or vertically into 2 groups and you get 4 in each group.
   8 ÷ 4 . . . Divide the columns and there are 2 in each column.

These cards offer a visual scaffold to those who need it, for as long as they need it. The goal is to eventually transition to the abstract, or simply numbers on a triangle. What you could do to support the transition is have the kids draw both on the front!

   Throughout the year, students should practice multiplication and division. As students are learning their facts and working on them, here is a set of quick quizzes that you can give to monitor progress (see Figures 4.58–4.64). Also remember to check out www.mathfactfluencyplayground.com for tons of division facts activities, games, and flashcards.

Dividing by 5 Quiz

Name:
Date:

$15 \div 5 =$	$20 \div 5$	Solve.
Model using equal groups.	Model using an array.	If 35 inches of yarn is cut into 5 inch pieces., how many pieces will it be cut into?
		Write the equation:
		Strategy: _____
		Answer: _____
Solve. During one month, Kevin ran a total of 40 miles. If he ran 5 miles each run, how many times did he run? Write the equation: Strategy: _____ Answer: _____	Solve. $30 \div \_\_\_\_ = 6$ $\_\_\_\_ \div 5 = 9$ $\_\_\_\_ = 25 \div 5$ $1 = \_\_\_ \div 5$	Make up your own problem dividing by 5 and solve.

$45 \div 5 =$

Model using a number line.

Circle how good you think you are at dividing by 5 facts!

Great	Good	Ok, still thinking

FIGURE 4.58 Dividing by 5 Quiz

Dividing by 10 Quiz

Name:
Date:

$30 \div 10 =$ Model with a drawing.	$40 \div 10 =$ Model using equal groups.	Solve. 70 marbles were evenly put into 10 bags. How many marbles were in each bag? Write the equation: Strategy: _____ Answer: _____
Solve. If Hannah practiced the piano every day for 10 minutes, how many days did she practice if she practiced for 50 minutes this week? Write the equation: Strategy: _____ Answer: _____	Solve. $20 \div$ _____ $= 2$ _____ $\div 10 = 7$ _____ $= 90 \div 10$ $0 =$ _____ $\div 10$	What is a strategy when dividing by 10? Explain with numbers, words, or pictures

$60 \div 10 =$
Model on a number line:

Circle how good you think you are at dividing by 10 facts!

Great　　　　　　　　　Good 　　　　　　　　　Ok, still thinking

FIGURE 4.59　Dividing by 10 Quiz

Dividing by 3 Quiz

Name:
Date:

$12 \div 3 =$ Model with a drawing.	$18 \div 3 =$ Model using an array.	Solve. Addison wanted to share 21 gum balls equally with two of her friends. How many gumballs can the three friends have? Write the equation: Strategy: _____ Answer: _____
Solve. A garden had 15 strawberry plants planted evenly in 3 rows. How many plants were in each row? Write the equation: Strategy: _____ Answer: _____	Solve. $6 \div$ _____ $= 2$ _____ $\div 3 = 6$ _____ $= 27 \div 3$ $1 =$ _____ $\div 3$	Make up your own problem dividing by 3 and solve.

$24 \div 3 =$

Model on a number line.

Circle how good you think you are at dividing by 3 facts!

Great   Good   Ok, still thinking

FIGURE 4.60 Dividing by 3 Quiz

Dividing by 6 Quiz

Name:
Date:

$24 \div 6 =$	$18 \div 6 =$	Solve.
Model using equal groups.	Model using an array.	An art teacher stored his paint brushes in containers that fit 6 brushes. If he had 30 paint brushes, how many containers did he need?
		Write the equation:
		Strategy: _____
		Answer: _____

Solve.	Solve.	What is a strategy when dividing by 6? Explain with numbers, words, or pictures
One apple pie recipe requires 6 apples. If a baker has 42 apples, how many pies can she make?	$48 \div 6 =$ _____	
Write the equation:	_____ $\div 6 = 4$	
Strategy: _____	_____ $= 36 \div 6$	
Answer: _____	$1 =$ _____ $\div 6$	

$54 \div 6 =$

Model using a numberline.

Circle how good you think you are at dividing by 6 facts!

Great   Good   Ok, still thinking

FIGURE 4.61 Dividing by 6 Quiz

Dividing by 4 Quiz

Name:

Date:

| $24 \div 4 =$

Model with a drawing. | $32 \div 4$ on the ten frame. 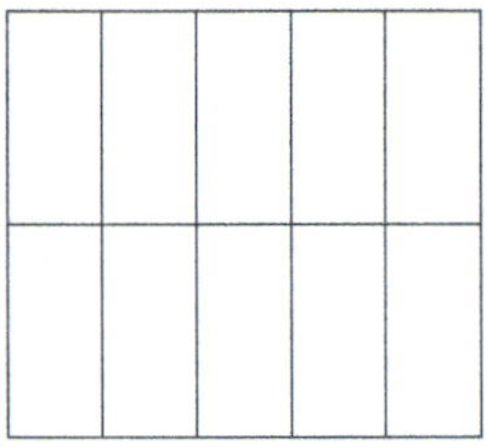 | Solve.
Matt had 16 treats for his four cats. If each cat received the same amount, how many treats did each cat get?

Write the equation:

Strategy: _____

Answer: _____ |
| Solve.
Joey spent $ 28 on baseball cards. If each pack cost $4, how many packs did he buy?

Write the equation:

Strategy: _____

Answer: _____ | Solve.

$32 \div 4 = $ _____

_____ $\div 4 = 5$

_____ $= 20 \div 4$

$0 = $ _____ $\div 4$ | What is a strategy when dividing by 4? Explain with numbers, words, or pictures |

$36 \div 4 =$

Model using a number line.

Circle how good you think you are at dividing by 4 facts!

Great   Good   Ok, still thinking

FIGURE 4.62   Dividing by 4 Quiz

Dividing by 8 Quiz

Name:
Date:

32 ÷ 8 = Model on a Numberline	16 ÷ 8 in equal groups Model using an array.	Solve. The physical education teacher divided her class into teams with 8 students on each team. If she had 24 students, how many teams did she have? Write the equation: Strategy: _____ Answer: _____
Solve. David slid over 64 beads on his rekenrek in 8 equal rows. How many beads were in each row? Write the equation: Strategy: _____ Answer: _____	Solve. 48 ÷ 8 = _____ _____ ÷ 8 = 9 _____ = 56 ÷ 8 5 = _____ ÷ 8	What is a strategy when dividing by 8? Explain with numbers, words, or pictures

24 ÷ 8 =

Model using a number line.

Circle how good you think you are at dividing by 8 facts!

Great   Good   Ok, still thinking

FIGURE 4.63   Dividing by 8 Quiz

Dividing by 9 Quiz		
Name: Date:		
$27 \div 9 =$ Model using equal groups.	$18 \div 9$ on the ten frame. Model using an array.	Solve. A rectangular garden measures 54 square feet. If the length of the garden is 9 ft, what is the width? Write the equation: Strategy: _____ Answer: _____
Solve. Charlie displayed his stamp collection in 9 equal rows. If he has 63 stamps, how many stamps were in each row? Write the equation: Strategy: _____ Answer: _____	Solve. $36 \div 9 =$ _____ _____ $\div 9 = 8$ _____ $= 18 \div 9$ $0 =$ _____ $\div 9$	What is a strategy when dividing by 9? Explain with numbers, words, or pictures
$81 \div 9 =$ Model using a number line.		
Circle how good you think you are at dividing by 9 facts! Great Good Ok, still thinking		

FIGURE 4.64 Dividing by 9 Quiz

KEY POINTS

- Learning trajectory of division
- Foundational facts
- Derived facts
- Goal setting

SUMMARY

Division can be tricky. We can't rush students to division. We need to make sure that they understand multiplication first. We need to give them plenty of opportunities working with manipulatives, acting out stories, drawing stories, and telling stories so that they own the ideas of division. Through distributed, intentional, engaging practice throughout the year, every child can learn division.

REFLECTION QUESTIONS

1. What stands out for you in this chapter?
2. What are your strengths with assessment and what are some challenges?
3. Can you state or draw a three-step action plan based on this chapter?

CALL TO ACTION

 1. Share your favorite "Aha!" moment about teaching division on social media to help spread the movement! #FDJH

 2. Take a photo of different ways that you are modeling basic division facts in your classroom and share it on social media to encourage other teachers to do it too! #FDJH

3. Get started with reasoning about division with this balance puzzle card game!

REFERENCES

Baroody, A. (2006, August). Why children have difficulties mastering the basic number combinations and how to help them. *Teaching Children Mathematics, 13*(1), 22–23.

Bay-Williams, J. M., & SanGiovanni, J. (2021). *Figuring out fluency in mathematics teaching and learning, grades K-8: Moving beyond basic facts and memorization.* Corwin.

Brendefur, J., Strother, S., Thiede, K., & Appleton, S. (2015, August). Developing multiplication fact fluency. *Advances in Social Sciences Research Journal, 2*(8), 142–154.

Carpenter, T. P., Fennema, E., Franke, M. L., Levi, L., & Empson, S. B. (2014). *Children's mathematics: Cognitively guided instruction* (2nd ed.). Heinemann.

Kling, G., & Bay-Williams, J. M. (2021, November). Eight unproductive practices in developing fact fluency. *Mathematics Teacher: Learning and Teaching PK-12, 114*(11), 830–838.

Kouba, V. L. (1989). Children's solution strategies for equivalent set multiplication and division word problems. *Journal for Research in Mathematics Education, 20*(2), 147–158.

Meg—The Herding Game. Retrieved December 29, 2022, from https://theteacherstudio.com/the-herding-game-introducing-division/

Mulligan, J., & Mitchelmore, M. (1997). Young children's intuitive models of multiplication and division. *Journal for Research in Mathematics Education, 28*(3), 309–330.

National Research Council, Mathematics Learning Study Committee. (2001). *Adding it up: Helping children learn mathematics* (Kilpatrick, J., Swafford, J., & Findell, B., Eds.). The National Academies Press.

Newton, N. (2018). *Daily math thinking routines in action: Distributed practices across the year.* Routledge.

CHAPTER 5

Multiplication and Division Problem Types

COGNITIVELY GUIDED INSTRUCTION—STARTING WITH STRUCTURE

When we think of facilitating fluency with basic facts, we typically think of giving students math facts as naked calculations and then observing how they think and reason as they determine answers. To make the experience even more meaningful and to support reasoning, it is powerful to situate the facts in contexts. We know that our approach with basic facts sets a foundation of flexible thinking that will follow our students as they encounter different sets of numbers. To support the conceptual development of the operations, it is powerful for students to have opportunities to explore story situations, preferably with real-life contexts that are relevant to their lives. This grounds the work with numbers so that students can grapple with them in a way that fosters sense making.

There has been a great deal of research on the effectiveness of schema-based word problem teaching and learning (Griffin & Jitendra, 2009; Murata, 2008; Fuchs et al., 2004; Willis & Fuson, 1988). In order for us to facilitate this journey, we must build our own understanding of the various story problem types so that we can be sure our students have exposure to them all. In this chapter, we will explore the various story problem types, model the structure of the story problems, and share some of our favorite ways for students to engage in the problem-solving experience.

One of the most influential books on the various problems types from which many state standards pull from is *Children's Mathematics: Cognitively Guided Instruction* (Carpenter et al., 2015). The researchers provided students with story situations without giving any directions and then observed how the children solved the problems, making notes of their actions, thinking, and reasoning. The authors state:

> One of the most useful ways of classifying word problems focuses on the types of action or relationships described in the problems. Research has shown this classification corresponds to the way that children think about the problems. As a result, this scheme distinguishes among problems that children solve differently and provides a way to identify the relative difficulty of various problems.
>
> (Carpenter et al., 2015)

We live by the Dr. Maya Angelou quote, "You do the best you can until you know better. When you know better, do better." We implore you to not do as we did by having key word

DOI: 10.4324/9781032614229-5

posters on your classroom walls or engaging in problem-solving steps that include highlighting numbers and underlining key words. There are many problematic issues with this practice. First, math practice standard #1 states that students are to "make sense of problems and persevere in solving them" (Council of Chief State School Officers, and National Governors' Association. Common Core State Standards Initiative. United States, 2022). When we have students isolate numbers and then look for a signal word to help them decide what operation to perform, we have prioritized getting the right answer over the value of engaging in thinking and reasoning about the story problem. Instead of suggesting that there is a trick that can help them get the answer and move on, we want students who are curious and willing to engage in thinking and reasoning, which is why the word "persevering" is so important. We know from both research and experience that in addition to fostering thinking and reasoning, this practice helps to develop positive math dispositions. Research has shown that students learn most when they are in a place of productive struggle (Boaler, 2015; SanGiovanni et al., 2020).

Another reason why using key words to determine an operation is problematic is that there are many times that the key word suggests an operation that is not appropriate for the given situation. Consider the problem, "Miguel collected 3 times as many pieces of sea glass as Rebecca. If Miguel had 9 pieces of sea glass, how many did Rebecca have?" If students have associated the "times as many" with multiplication, they may choose to multiply the 9 by 3 to determine that Rebecca had 27 pieces of sea glass. But this isn't reasonable because Miguel only has 9 pieces, and he has more than Rebecca. Students will calculate an answer but have no idea whether it is reasonable or not because thinking and reasoning are absent from the process.

A final reason why using key words to determine which operation to use is problematic is because it doesn't allow for students to think flexibly and work with numbers in ways that make the most sense for their brains. If the structure of a story problem is such that one of the factors is unknown, students have the choice of using any of the four operations to solve the problem. Granted, some strategies may be more efficient than others, but it is important that students develop agency when learning mathematics. This is a critical piece in developing their math identities and discovering connections between and among operations on their own road to efficiency. It is our role to be facilitators of this journey and help them move toward more efficient and effective strategies as they calculate their answers.

Students need the freedom to engage in sense making. One way to this is by visualizing the structure of the story problem and then, given the structure and numbers involved, using strategies based on properties and relationships that are most efficient for the situation. In this way, we are developing young mathematicians who are thinkers and reasoners. This breeds the confidence to try problems that may look more difficult and develops positive math dispositions that empower them to believe they can figure problems out.

Some of the most impactful work we can do with students is to help them create models of the structure of problem types when they encounter story problems. Since these problem types will reappear across various grade levels, understanding the structure will help students to understand how to enter the problems and how to develop a plan to solve them regardless of the operations involved. As an example, the same problem types seen in single-step problems are later combined together to create two- and multi-step problems. It is imperative, then, that we explore these various situations with students using concrete, pictorial, and abstract representations. These experiences will build their foundational understanding of the operations of multiplication and division, as well as the inverse relationship between multiplication and division, supporting a strong foundation on which to build beyond basic facts.

EQUAL GROUP AND ARRAY SITUATIONS INVOLVING MULTIPLICATION

When students are first learning about situations that involve multiplication within 100, typically in grade 3, it is suggested by Carpenter et al. (2015) that we begin with equal groups and arrays. For equal groups, we know the *number of groups* and the *amount in each of the groups*, so we are looking for the total amount. For arrays, there are distinct objects arranged in equal-sized *rows* and *columns*, and we are determining the total amount. One of the most powerful models we can use with both multiplication and division is the area model, which is the measurement version of this array problem type. Students will know the lengths of the sides and then can determine the area within the rectangle. Within these basic problem types, students can explore different situations and use different representations along the CRA continuum to model them (see Figure 5.1).

Problem Types	Equal Groups	Arrays	Area	Rate	Price
Multiplication	Kylynn had 4 containers of toy trucks. If each container had 3 toy trucks, how many does she have in all?	Denise had 3 rows of rhubarb plants. Each row had 5 plants. How many rhubarb plants did she have?	A rectangular garden measured 6 ft by 7 ft. What was the area of the garden?	Jen reads 2 books each week. How many books will she read in 3 weeks?	Ice cream cones cost $4. If Sue buys herself and her friend an ice cream cone, how much will she spend?
Bar Diagram Modeling Problem	? / 3 3 3 3	5 / 5 / 5 ?	7ft. 6ft.	? / 2 2 2	? / $4 $4
What are we looking for? Where is X?	In this type of story we are looking for the total amount of toy trucks in all the groups.	In this type of story we are looking for the total amount of all the rows of rhubarb plants..	In this type of story we are looking for the area.	In this type of problem we are looking for the total amount of books read..	In this type of problem we are looking for the total price.
Algebraic Sentence	3 + 3 + 3 + 3 = 12 4 x 3 = 12	5 + 5 + 5 = 15 3 x 5 = 15	6 + 6 + 6 + 6 + 6 + 6 + 6 = 42 7 + 7 + 7 + 7 + 7 + 7 = 42 6 x 7 = 42	2 + 2 + 2 = 6 3 x 2 - 6	4 + 4 = 8 4 x 2 = 8
Strategies to Solve	Repeated Addition Multiply	Repeated Addition Multiply	Repeated addition Multiply	Repeated Addition Multiply	Repeated Addition Multiply
Answer	Kylynn had 12 toy trucks	Denise had 15 rhubarb plants.	The garden has an area of 42 square feet.	Jen reads 6 books in 3 weeks.	Sue spent $8.

FIGURE 5.1 Multiplication Problem Types and Models

(Adapted from Newton 2018)

Research has shown that tape diagram models are very powerful to help students visualize the structure of problems (She & Harrington, 2022; Ding, 2018; Murata, 2008). As She and Harrington (2022) state, "When an accurate tape diagram is created, it manifests itself for a visual display of relationships, and students are able to setup operations with numbers and symbols based on where the question mark(s) are in each diagram." Throughout our book, we have emphasized the importance of concrete, representational, and abstract representations of basic facts, and this is equally as important when exploring the underlying structures of the problem types. Some of our favorite manipulatives are beaded number lines, one-inch tiles or cubes, and Cuisenaire® rods. Let's take a look at an example of an equal group and an array story problem using this variety of concrete, representational, and abstract representations. Using Brainingcamp's virtual manipulatives (www.brainingcamp.com), we can demonstrate concrete and pictorial tape diagram models of the structure of the problem types. At the same time, we can capture and record the abstract representation of the thinking of the students as they solve (see Figures 5.2–5.4).

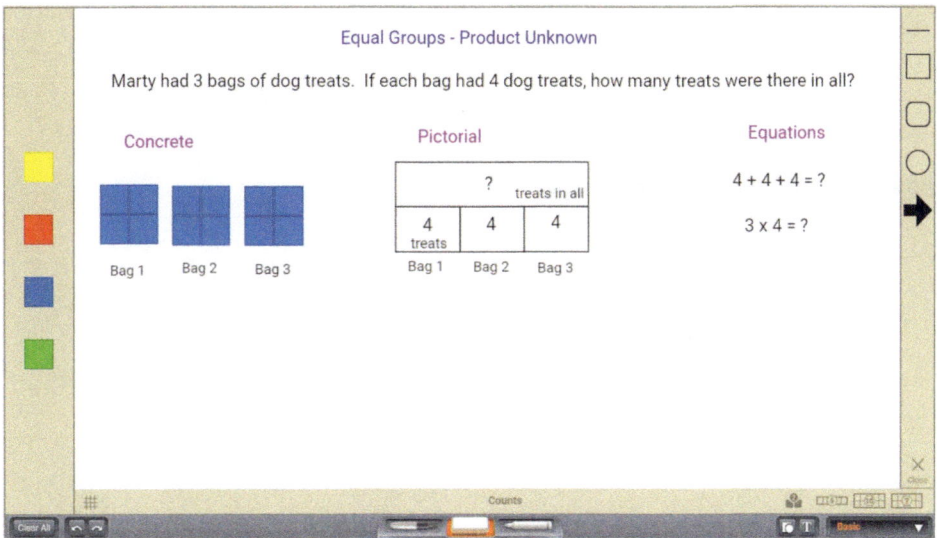

FIGURE 5.2 Equal Group Problem Type

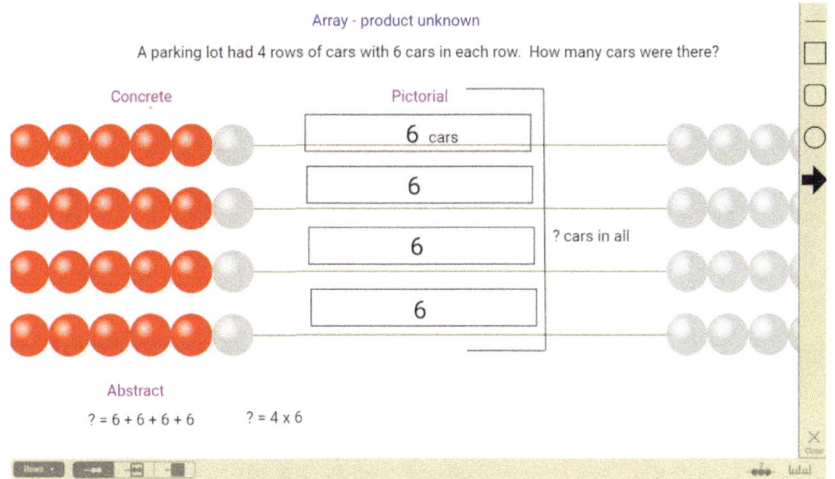

FIGURE 5.3 Array Problem Type

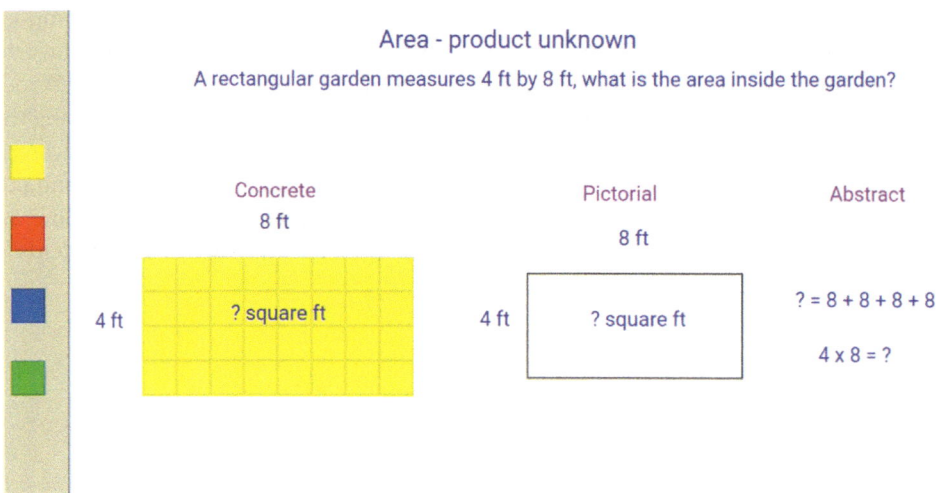

Area - product unknown

A rectangular garden measures 4 ft by 8 ft, what is the area inside the garden?

Concrete

Pictorial

Abstract

8 ft

8 ft

4 ft   ? square ft

4 ft   ? square ft

$? = 8 + 8 + 8 + 8$

$4 \times 8 = ?$

FIGURE 5.4   Area Measurement Problem

EQUAL GROUP AND ARRAY SITUATIONS INVOLVING DIVISION

There are several different types of problems that may be solved using division, although students don't necessarily need to do division to calculate the answers (see Figure 5.5). Typically, in grade 3, students will explore equal group stories, only this time, rather than the total amount being unknown, the *number of groups* or the *number in each group* will be unknown. In addition, students explore the array problem type where the total amount is known, but the unknown is the number of either rows or columns. Area problems would require students to solve for one of the sides of a rectangle given the area and the length of one side. Figure 5.5 outlines examples of these problem types.

Problem Types	Amount in Each Equal Group	Amount of Equal Groups	Arrays How many In each group?	Arrays How many groups?	Area	Rate	Price
Division	Miriam wanted to share 12 gumballs equally with two of her friends. How many gumballs can each person get?	A gardener had 28 flowers. If 7 flowers fit in a vase, how many vases did she use?	A baker baked a pan of 24 cookies placed in 4 equal rows How many rows were there?	A baker baked a pan of 12 cookies with 4 cookies in each row. How many rows of cookies were there?	Juan had a playroom that had an area of 72 square feet. If the width of the room was 8 ft what was the length?	Jaritza practiced piano for 10 minutes each session. If her goal is to practice for 60 minutes each week, how many times will she need to practice?	Cathery spent $14 on 2 magazines. If they cost the same amount, how much did each magazine cost?
Bar Diagram Modeling Problem	12 [? ? ?]	28 [7 → 7] ?	? [? ? ? ?] 24	12 [4] [4]	? ft [8 ft] Area = 72 sq ft	60 [10 →]	14 [? ?]
What are we looking for? Where is X?	In this type of story we are looking for the amount in each group.	In this type of story we are looking for the amount of groups.	In this type of story we are looking for the amount in each row.	In this type of story we are looking for the amount of rows.	In this type of story we are looking for one of the sides.	In this type of problem we are looking for the rate.	In this type of problem we are looking for the price.
Algebraic Sentence	3+3+3+3=12 12-3-3-3-3=0 3 x 4 = 12 12÷3 = 3	7+7+7+7=28 28-7-7-7-7=0 7x4=28 28÷7 =4	4+4+4+4+4 +4=24 24-4-4-4-4-4 -4=0 4x6=24 24÷4 = 6	4+4+4=12 12-4-4-4=12 4 x 3=12 12÷4 = 3	8+8+8+8+8+8 +8+8 = 72 72-8-8-8-8-8 -8-8-8=0 8 x 9 = 72 72 ÷ 8 = ?	10+10+10+10 +10+10=60 60-10-10-10 -10-10-10=0 10 x 6 =60 60 ÷ 10 = 6	2+2+2+2+2 +2+2=14 14-2-2-2-2-2 -2-2=0 2 x7=14 14 ÷ 2 =7
Strategies to Solve	Repeated addition Repeated subtractio Think Multiplicatio Division	Repeated addition Repeated subtracti Think Multiplicatic Division	Repeated additior Repeated subtraci Think Multiplicat Division	Repeated additior Repeated subtract Think Multiplicat Division	Repeated additior Repeated subtract Think Multiplicat Division	Repeated additic Repeated subtra Think Multiplica Division	Repeated addition Repeated subtracti Think Multiplicatic Division
Answer	Each person may have 4 gumballs.	The gardener used 4 vases.	There are 6 cookies in each row.	There were 3 rows of cookies.	The length was 9 ft long.	It will take her 6 practice sessions.	Each magazine cost $7.

FIGURE 5.5 Division Problem Types and Models

(Adapted from Newton 2018)

Let's examine the structure of the equal group situation when the number of groups is known, but the amount in each group is unknown. This is called partitive division or fair sharing. Students take the amount of the divisor from the total and then equally distribute the value to each of the groups, beginning with one to each group but eventually using larger and larger groupings that may include place value understanding as the number sets become larger. The following figure is an example of this type of problem. Imagine a student taking 3 from the 15; giving 1 to Cathy, 1 to Tara, and 1 to Ann; and then taking another 3 and doing the same thing until there are none remaining (see Figure 5.6).

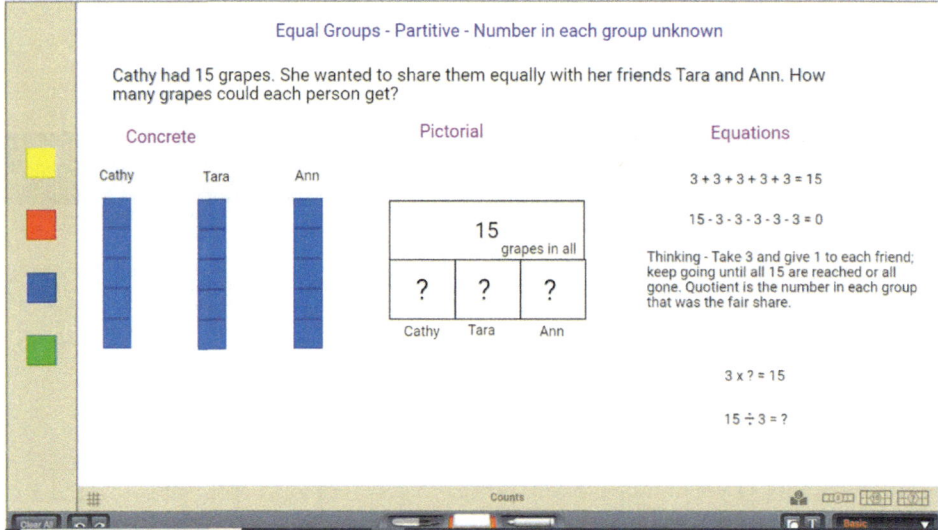

FIGURE 5.6 Equal Groups Number in Each Group Unknown

The last version of the equal group problem type is when the *size of the groups* is known, but the *number of groups* is unknown. This is referred to as quotative or measurement division. Students think of the amount of the divisor and rather than sharing it out between groups, and they know that the amount makes one group. Then, taking another group of that size would be another group, and so on. In this way, they are determining how many groups of that size can fit in the total. Again, students have the freedom to think about repeated addition, repeated subtraction, thinking multiplication, or thinking division. Students access whatever makes sense to them at the time depending on where they are with strategies and operations relative to the numbers are that are involved. Here is an example of this problem type. Imagine a student taking 3 and making one piece, taking another 3 and that being another piece, and so on (see Figure 5.7).

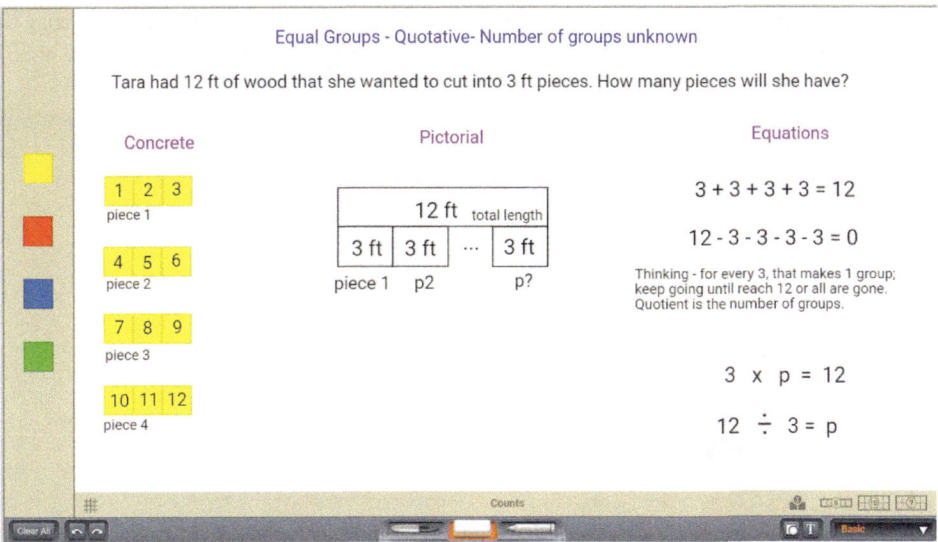

FIGURE 5.7   Equal Groups Number of Groups Unknown

Array situations have physical objects arranged in rows and columns. Division can be used when either the number of rows *or* the number of columns is unknown. Thinking about when the number in each row is unknown mirrors partitive or fair-share thinking, where the quantity of the divisor is taken and distributed to each of the rows. This is continued until the total amount is gone. This will become more sophisticated as students remove groups that are multiples of the divisor and continue to share equally among the groups (see Figure 5.8).

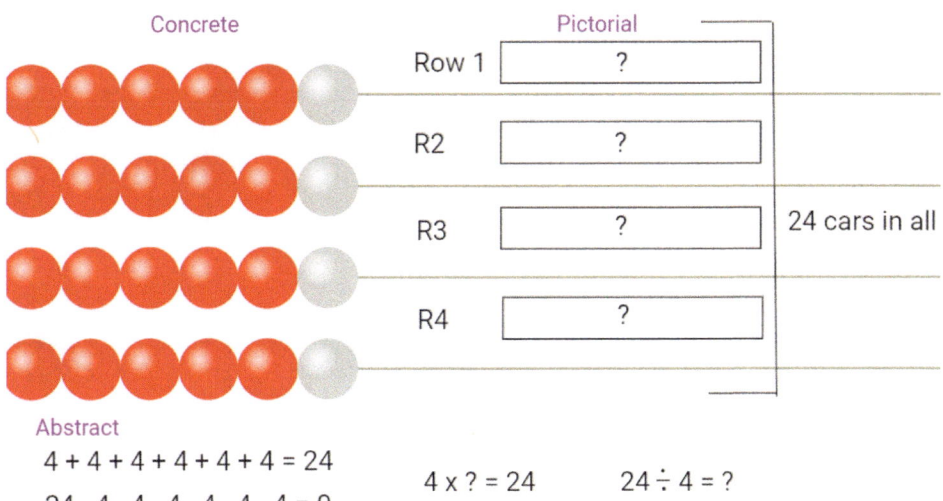

FIGURE 5.8   Array Number in Each Row Unknown

When an array situation involves determining the number of rows, this mirrors the quotative interpretation of division since we can remove the amount in the divisor that becomes one row and so on (see Figure 5.9).

Array - # of rows unknown

A parking lot had 24 cars parked 6 cars to each row. How many rows of cars were there?

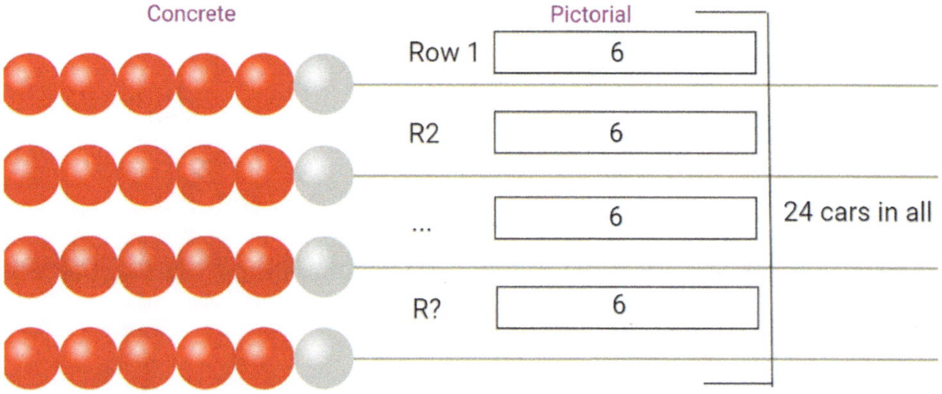

Concrete   Pictorial

Row 1 | 6 |

R2 | 6 |

... | 6 | 24 cars in all

R? | 6 |

Abstract

6 + 6 + 6 + 6 = 24   0 = 24 - 6 - 6 - 6 - 6   6 x ? = 24   24 ÷ 6 = ?

FIGURE 5.9 Array Number of Rows Unknown

Situations involving the area of a rectangular shape involve division when we know the area and one of the side lengths and need to determine the length of the other side. The thinking with this situation is neither partitive nor quotative. In this instance, students can think about partial products that make up the total if they are unable to recall from memory the division fact (see Figure 5.10).

Area model - side length unknown

A rectangular garden has an area of 48 square ft. If the width measures 6 ft, what is the length?

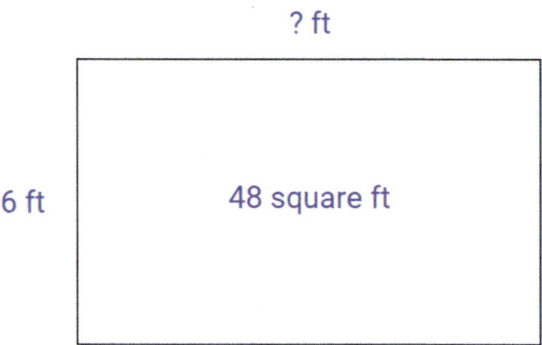

? ft

6 ft   48 square ft

FIGURE 5.10 Area Model Side Length Unknown

If we were to ask you to compare 3 to 12, what would you say? Would you say that 12 is 9 more than 3 or would you say 12 is 4 times as many as 3? Therein lies the difference between additive comparison and multiplicative comparison. Within a multiplicative comparison story problem, there is a larger amount and a smaller amount, and the relationship between the two is that the larger amount is *a certain number of times* bigger than the smaller. Since the unknown could be at any of those three spots, there are three different versions of multiplicative comparison problems (see Figures 5.11–5.15). If students have been relying solely on key words to solve problems, one version they will find particularly difficult is when we are looking for the smaller part. Consider this example: "Kylynn has 27 marbles; she has 3 times as many as her friend Riley. How many marbles does Riley have?" This is a tricky problem for students because they want to multiply instead of divide want they see the word *times* and want to solve it without thinking about the entire problem.

Make sure that you emphasize the difference between additive comparison problems and multiplicative comparison problems. Illustrative mathematics has a really good snake problem where they discuss measuring different sized snakes that grow the same amount. They ask the students to explain if the snakes grew the same amount and expect students to be able to discuss that there are a few ways to look at it. Both additively and multiplicatively. ***See Illustrative Mathematics Snake Problem 1 & 2
https://www.illustrativemathematics.org/content-standards/tasks/356

https://www.illustrativemathematics.org/content-standards/tasks/357

Figure 5.11

FIGURE 5.11   Additive vs Multiplicative

(Adapted from Newton 2017)

Problem Types	Bigger Part Unknown	Smaller Part Unknown	Difference Unknown (Number As Times As Many)
Compare	Hamilton had 4 dog bones. Marty had 2 times as many bones as Hamilton did. How many bones did Marty have?	Justina had 16 rings which was twice as many as her sister Bianca. How many does Bianca have?	Hannah owns 7 games. Matthew owns 21 games. How many times as many games does Matthew have?
Bar Diagram Modeling Problem	4 / 4 4 ?	? ? 16 / ?	7 / 7 → 21
What are we looking for? Where is X?	In this type of story we are comparing two amounts. We know the smaller amount and we have to find the larger amount.	In this type of story we are comparing two amounts. We are looking for the smaller part which is unknown.	In this type of story we are comparing two amounts. We know both parts but we are looking for *how many times as many* the larger part is than the smaller.
Algebraic Sentence	4 x 2 = 8	16 ÷ 2 = 8	7 x 3 = 21 or 21 ÷ 7 = 3
Strategies to Solve	Multiply Repeated Addition	Think multiplication Divide	Think multiplication Divide
Answer	Marty had 8 dog bones.	Bianca had 8 rings.	Matthew had 3 times as many games as Hannah.

FIGURE 5.12 Multiplicative Comparison Types

In this section are examples of the multiplicative comparison problem types using Brainingcamp (www.brainingcamp.com). We can use this tool to demonstrate concrete and pictorial models of the structure of the problem types and the abstract representation of the thinking of the students as they solve. In Figure 5.8, imagine a student taking the amount of the smaller group, in this case the 4 books, and then repeating it 3 times to then determine the amount that Sue must have. Students may begin by skip counting, but we want to encourage them to think multiplicatively and employ more efficient strategies based on number relationships.

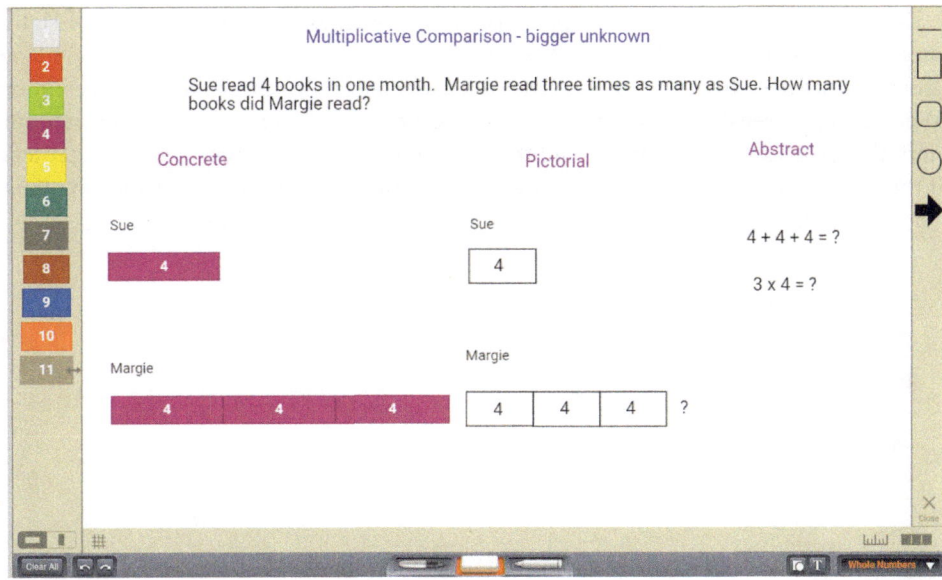

FIGURE 5.13   Multiplicative Comparison Bigger Unknown

Division may be used when one of the factors is unknown. As seen in Figure 5.9, we will know the larger group amount and the number of times larger the bigger is than the smaller. If students are using additive reasoning, there may be some guess and check for what number can be repeated three times to make the 15, but if they think about it as $15 \div 3 = ?$ or as $3 \times ? = 15$, they can utilize multiplicative reasoning.

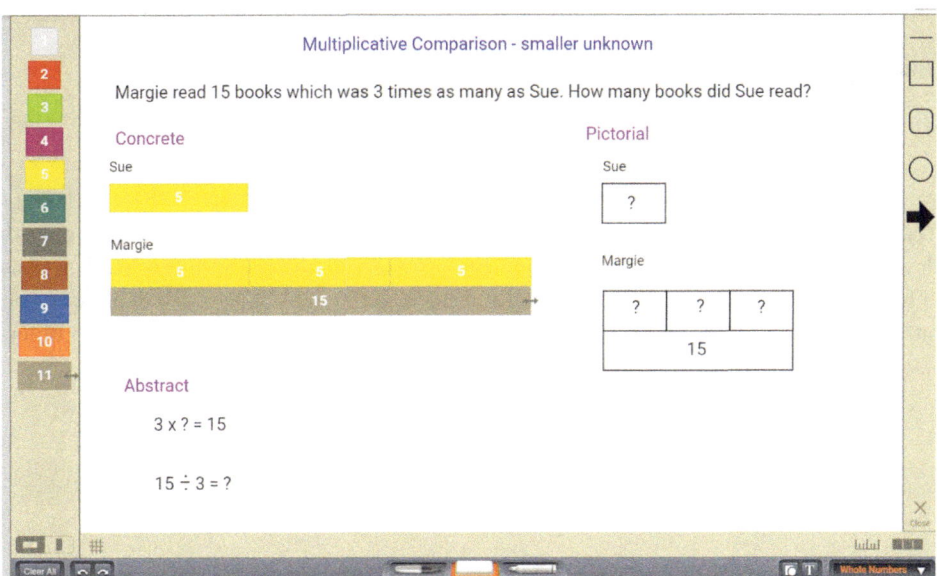

FIGURE 5.14   Multiplicative Comparison Smaller Unknown

In this final version of the multiplicative comparison story type (See Figure 5.14), the multiplier is unknown. We know the amount of the small group and the amount of the large group, but we need to determine how many of the small groups are within the larger group. When using Cuisenaire® rods, students may take a rod the size of the small group, repeat it until they reach the total amount, and then determine how many rods made up that total if they are thinking additively. As they transition to multiplicative reasoning, they may consider a division expression or think of the situation as a missing factor problem.

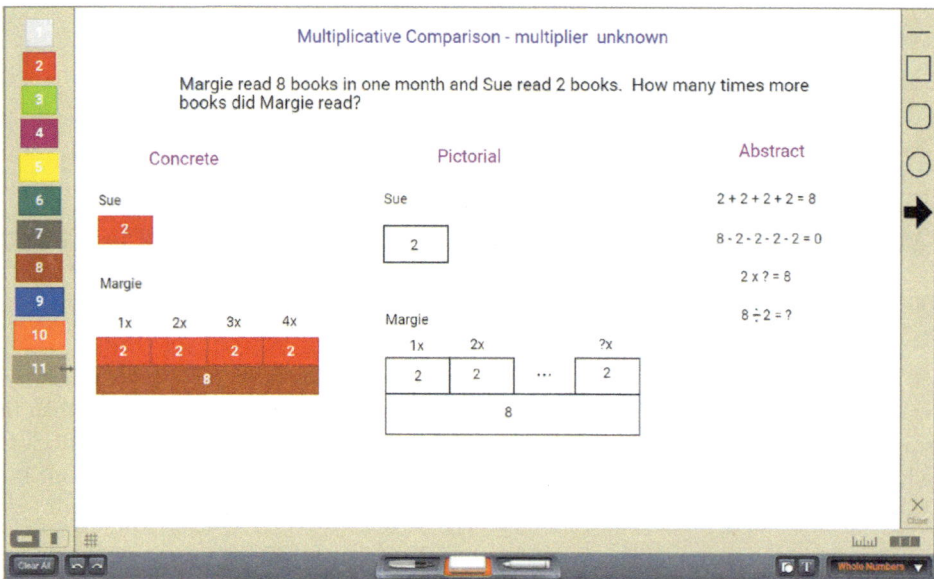

FIGURE 5.15   Two-Step Problems

Two-step and multi-step problems are combinations of these single-step problem types for all four operations. They can range from very easy to very difficult depending on the level of difficulty of each of the problem types involved. Making sure that students have a solid foundational understanding of all the single-step problem types with all four operations is essential. Here are some possibilities of two-step and multi-step story problems (see Figures 5.16 and 5.17).

Same Operation	Different Operations	Comparison	Mixed Levels	Mixed Levels of Harder Versions
A bakery placed 3 rows of muffins with 6 muffins in each row into a container. If they had 7 of those containers, how many muffins were there in all?	Marty bought 4 packages of stickers that had 18 stickers in each package. If he then gave a sticker to each of his 24 classmates, how many did he have left?	Dan ran ¾ mile and Chris ran three times as far as Dan. How far did they run all together?	A florist received a shipment of 7 dozen roses. If he needed to make 6 vases of floral arrangements using all the roses equally in each vase, how many roses were there in each vase?	Kathryn needed to purchase furniture for her apartment. The couch cost $1275. The dining room table cost 4 times as much as the chairs. If she spent $3475, how much did she spend on the table and the chairs? How much did she spend on each of the 4 chairs?

FIGURE 5.16

(Cited in Newton 2017; Adapted from CCSSM Progressions)

Multi-Step Problems	
Denise read 5 books. Jen read 2 times as many books as Denise. How many did they read altogether? If Jen then reads 5 more books, how many times more books will she have read than Denise?	It cost $16 for children and $24 for adults to get into the science museum. Bag lunches cost $8.50. If a family with 2 adults and 4 children spent the day at the museum and each person got a bag lunch, how much money would they have left if they brought $200?

FIGURE 5.17

HOW DO WE EXPLORE PROBLEM SOLVING WITH STUDENTS?

As discussed throughout this chapter, there are many different problem types that involve situations that can be solved using multiplication or division. While students don't need to be able to name the types specifically, they should be able to discuss the situation while thinking and reasoning in order to visualize the structure and determine a solution. It is the structure of the story situations that teachers should emphasize and that students should model. They should be able to describe what is happening in the story, what the quantities represent, and what are they looking for. Students should understand whether or not they are looking for the amount in each group or if they are looking for the amount of groups. They should think about whether the prompt is a single-step, two-step, or even multi-step problem. This helps students to unpack problems and make a plan for solving them. If students know what they are looking for, they are much more likely to find it!

THREE-READ PROTOCOL

One of the most powerful ways to explore problem solving with students is using a three-read protocol. There has been some amazing work done by the San Francisco Unified School District implementing this protocol. You can find lots of free resources on their district site www.sfusdmath.org/3-read-protocol.html. Additionally, there are some wonderful videos on YouTube that show Duane Habecker leading the three-read protocol with kindergarten students (https://youtu.be/hRSUSyUu0CA), 2nd graders (https://youtu.be/r5i90Rao0rw), and 5th graders (https://youtu.be/NpVR7cgzxdY). Essentially, you use any story problem, remove the last question, and then display it for the students. During the first read, the teacher will read the problem orally to the class and then have students discuss what is happening in the story problem. For the second read, the teacher and students do a choral read and then the discussion focuses on the quantities (explicit or implied) that are in the story. The teacher records this information on a chart. For the third read, the teacher and students again read the situation together orally. Then, students generate questions that could be asked given the information in the situation while the teacher records these questions.

Once a variety of questions have been written that could be solved using the given information, there are many options for how to proceed. One option is that the teacher selects one of the questions for everyone in the class to solve. Typically, this would be the question that was originally removed from the story problem. Students could then work in small groups or with a partner to solve the problem, making sure to use models that make their thinking visible to others. Another option would be for students to answer their choice of any of the questions. Again, they may work in small groups or with partners to solve the problem and make their thinking visible using concrete, pictorial, and/or abstract ways. When students choose their own question to solve, there will be various correct solutions, thus providing an opportunity for students to enter the problem-solving activity regardless of where they are on the learning trajectory. The power of this routine is that not only are students given ample opportunity to discuss the story situation before they solve it but they also share with others how they chose to solve it in efficient and effective ways.

Here is an example of a story problem that could be given to students that offers the possibility of various solutions.

> *Ben went to an apple orchard in the fall. He saw that a bushel of Cortland apples cost $ 4.50, a bushel of Macintosh apples cost $ 5.25, apple pies cost $ 6.50, and apple cider donuts cost $ 3.75 for a half dozen. Ben brought $ 20 to the orchard.*

Using this information, students could write questions that could be solved. The questions could be single-step such as determining how much money it would cost for two bushels of Cortland apples, or multi-step with determining the change Ben would get back if he were to purchase a few of the items. The possibilities are endless! Each student can approach this situation at their own readiness levels and have the opportunity to share their reasoning with others.

NUMBERLESS WORD PROBLEMS

Another powerful problem-solving routine is called numberless word problems. Brian Bushart's website (https://bstockus.wordpress.com/numberless-word-problems/) explains the evolution of this routine in his district and includes many slide decks ready for you to use with your students. Too often, when students are given a word problem to solve, they just want to grab the numbers and calculate an answer (usually determined by a key word). In this routine, the problem-solving process is slowed down by only revealing one sentence or one piece of information at a time. In this way, students get the chance to discuss the information they are receiving, determine how that relates to what they already know, and think of possible questions that could be asked before the final question is revealed. This provides plenty of opportunity for students to discuss what is happening in the story as well as the underlying structure of the story problem before they jump in to solve it.

The following figure is an example that shows how each piece of information is revealed slowly (see Figure 5.18). After each bit of information is revealed, students are encouraged to discuss what new information is known and how it affects what they know about the situation. Once all of the information is given, students brainstorm questions that could be answered using the given information.

A squirrel collected some acorns in the morning. He collected some more acorns in the afternoon.

A squirrel collected 72 acorns in the morning. He collected some more acorns in the afternoon.

A squirrel collected 72 acorns in the morning. He collected 59 acorns in the afternoon.

How many acorns did the squirrel collect that day?

How many fewer acorns did he collect in the afternoon than the morning?

How many more acorns would the squirrel have to pick to have picked the same as in the morning?

FIGURE 5.18

OPEN WORD PROBLEMS

Typically, when we think of problem solving, we imagine a story problem that will result in one correct answer. In the real world, however, there are often multiple solutions to problems. For this reason, we want to be sure to provide students with opportunities to grapple with situations that have many correct answers. Open word problems, such as this next one, support this type of thinking.

> *Jen shared 18 stickers equally among her friends. She had none left. How many friends could she have shared them with?*

After reading this problem, students could be encouraged to show their thinking in concrete, pictorial, and abstract ways and perhaps even share more than one possible solution. This routine helps to develop our students as flexible, confident problem solvers. As an added bonus, it sends the message that math is not just about finding one correct answer.

CONTEXTUALIZING AND DECONTEXTUALIZING

So far, we have discussed how students can decontextualize story problems by taking the known situations and then modeling their solutions. Having students decontextualize is very common in textbooks, worksheets, and task cards, but how often do we ask students to create contexts for problems? In this scenario, students are given a model and asked to create a context to match it. When students engage in contextualizing, it is helpful if they can relate it to a real-life situation that they are familiar with. As seen in Figures 5.19 and 5.20, you can share models that are either concrete, representational, or abstract, depending on the student. Consider these models and what story situations your students might write:

The situations could be expressed in images:

FIGURE 5.19

FIGURE 5.20

Or bar models:

36			
N	n	N	n

Once students have brainstormed contexts, the class could then model the structure of the story problems, possibly without even solving them. By using these routines, students will deepen their understanding of the underlying structures of the story problems. In addition, they will be free to solve them in ways that make sense to them no matter where they are on their math journeys. Such experiences help us to develop confident and flexible problem solvers.

COMBINATIONS

Combination problems are wonderful thinking problems for students. We don't tend to spend as much time doing these problems, but that doesn't mean that they aren't valuable! As always, we want to engage in a cycle of concrete, representational, and abstract with these problems to build understanding.

> *Story: Grace went to the ice cream parlor. She got 1 scoop with 1 topping. There were 2 kinds of ice cream: chocolate and strawberry. There were 3 kinds of topping choices: plain, sprinkles, or dipped in chocolate.*

In this poster, you can see 3 parts: concrete, representational, and abstract explorations of the ice cream problem.

Concrete Representation

Give the students construction paper ice cream cones, scoops, and toppings. They have to figure out all of the combinations that can be made. It is important to do it concretely rather than just have students color in options.

Pictorial Representation: Make a Tree Diagram to Show the Different Choices She Could Have Had

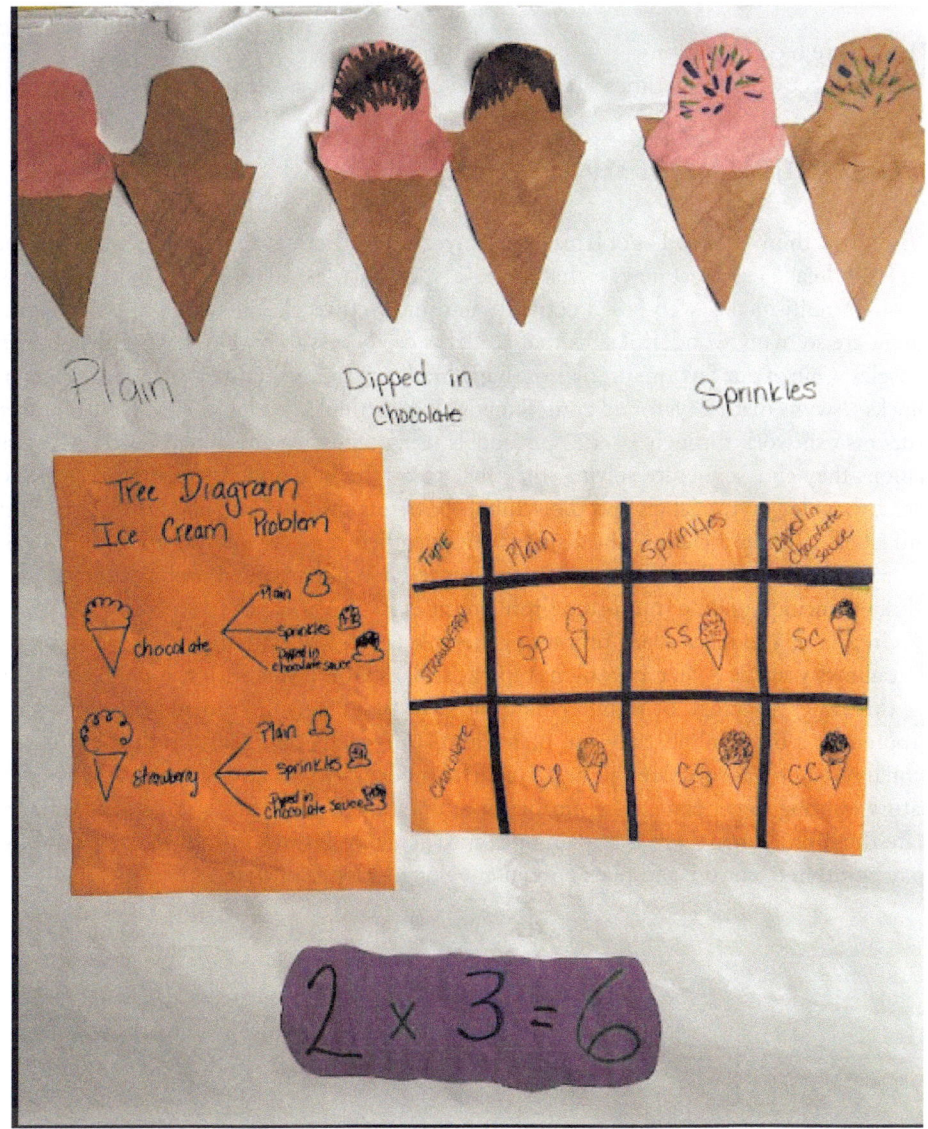

FIGURE 5.21 Combinations

Abstract Representation

$3 \times 2 = 6$

There were 6 combinations.

Do this activity several times. Have the students help you generate several different ideas.

WORD PROBLEMS IN THE 21ST CENTURY

When you think of word problems, you may think of boring, unrelatable scenarios that students dread! The great news is that in the 21st century, "word" problems have come alive through multi-media, web-based activities and interactives that may live in workstations. There are so many wonderful sites that students can access to build their problem-solving muscles! One of our favorite problem-solving modeling sites is Math Playground's Thinking Blocks (www.mathplayground.com/ThinkingBlocks/thinking_blocks_start.html), where students can work through a series of models using tape diagrams based on the story situations they choose. Before solving, they first move the bar models to appropriate places. Next, they drag in the labels and the symbol for the unknown. Once the problem is set up and labeled, students can use the visual to support their thinking in order to solve. This site is very versatile and allows teachers or students to choose the type of problem they wish to work on, including some multi-step problem types.

Greg Tang also has a word problem generator (https://tangmath.com/wordproblems) that allows you can select a problem type as well as where you wish the unknown to be. At this time, the range of numbers is limited to basic facts. If students are solving one problem at a time online, they can press "hint" to reveal a tape diagram of the situation, which research has suggested to be a powerful model (She & Harrington, 2022). For educators wanting to generate many problems at one time for a workstation and to differentiate these problems, there is also an option to generate 10 word problems at a time that may be printed, cut up, and placed in an envelope for workstations.

Picture Books/ Stories	Paper and Virtual Tools	Internet Word Problem Sites	Online Resources
Use picture books as a launch into different problem contexts. There is a great book called *Tall Tale Math* specifically for grades 3–5 that mathematizes tall tales.	Use a variety of physical and virtual manipulatives to solve problems such as: Cuisenaire® rods 10 frames Rekenreks Number lines For virtual manipulative sites: Brainingcamp (www. brainingcamp.com) Math Learning Center (www. mathlearningcenter.org/ apps) Didax (www.didax.com/math/ virtual-manipulatives. html)	Numberless word problems: https:// bstockus.wordpress. com/numberless-word-problems/ Three-read protocol: www.sfusdmath.org/3-read-protocol.html Math Playground's Thinking Blocks: www.mathplayground. com/thinkingblocks.html Greg Tang's word problem generator https://tangmath.com/ wordproblems	South Dakota word problem resources: https://sddial.k12. sd.us/esa/grants/ sdcounts/ Illustrative Math www. illustrativemathematics. org/ Inside Mathematics www.insidemathematics. org/

FIGURE 5.22

KEY POINTS

- Cognitively Guided Instruction (CGI) provides a framework for teaching word problems (Carpenter et al., 2015).
- The emphasis should be on the problem types and structure rather than keywords.
- There are 3 major categories for multiplication and division.
- There are varying levels of multi-step problems.
- There are many 21st-century tools to engage students in problem solving.

SUMMARY

Research has shown that schema-based problem solving is a powerful way to provide access to problem solving for all students (Carpenter et al., 2015). For many teachers, exposure to the research on the various problem types wasn't included in teacher training programs. Learning about the types and practicing visualizing the structure of the problems is important for us before we embark on exploring them with our students. During the problem-solving process, students should focus both on modeling the structure of the story situation and using computational strategies that make sense to them as they determine reasonable answers and persevere to solve. The role of educators is to listen to the thinking of the students and help guide students toward more and more sophisticated and efficient strategies.

There are many engaging and free resources to explore problem solving that offer the differentiated practice that students need to support their conceptual understanding of the operations of multiplication and division. When we begin this journey with basic facts, the students then have a foundation of thinking and conceptual understanding that will follow them as they explore multiplication and division with larger numbers, fractions, and decimals. A powerful journey indeed!

REFLECTION QUESTIONS

1. Do you presently teach with the word problem types in mind?
2. Do you presently get students to think about the steps involved in the word problem? If so, how do you currently do it? If not, how might this help?
3. Do you use technology to teach and practice word problems?
4. What are your biggest takeaways from this chapter?

CALL TO ACTION

 1. Share your favorite "Aha!" moment about teaching multiplication and division word problems on social media to help spread the movement! #FDJH

 2. Take a photo of different ways that you are teaching multiplication and division word problems in your classroom and share it on social media to encourage other teachers to do it too! #FDJH

3. Get started reasoning with robot mats. You can tell both multiplication and division problems.

REFERENCES

Boaler, J., Williams, C., & Confer, A. (2015). Fluency without fear: Research evidence on the best ways to learn math facts. youcubed@Standford University. Retrieved on May 12 , 2019 from https://www.youcubed.org/evidence/fluency-without-fear/

Carpenter, T. P., Fennema, E., Franke, M. L., Levi, L., & Empson, S. B. (2015). *Children's mathematics: Cognitively guided instruction (CGI)* (2nd ed.). Heinemann.

Council of Chief State School Officers, and National Governors' Association. Common Core State Standards Initiative. United States (2022). Web Archive

Ding, M. (2018). Modeling with tape diagrams. *Teaching Children Mathematics. NCTM, 25*(3), 158–165.

Fuchs, L. S., Fuchs, D., Finelli, R., Courey, S. J., & Hamlett, C. L. (2004). Expanding schema-based transfer instruction to help third graders solve real-life mathematical problems. *American Educational Research Journal, 41,* 419–445.

Griffin, C. C., & Jitendra, A. K. (2009). Word problem-solving instruction in inclusive third-grade classrooms. *The Journal of Educational Research, 102,* 187–201.

Murata, A. (2008). Mathematics teaching and learning as a mediating process: The case of tape diagrams. *Mathematical Thinking and Learning, 10*(4), 374–406, https://doi.org/10.1080/10986060802291642

Newton, N. (2017). *Math problem solving in action: Getting students to love problem solving grades 3–5.* Routledge.

SanGiovanni, J., Katt, S., & Dykema, K. (2020). *Productive math struggle: A 6-point action plan for fostering perseverance.* Corwin.

She, X., & Harrington, T. (2022). Teaching word-problem solving through tape diagrams. *Mathematics Teacher: Learning & Teaching. NCTM, 115*(3), 170–182.

Willis, G. B., & Fuson, K. C. (1988). Teaching children to use schematic drawings to solve addition and subtraction word problems. *Journal of Educational Psychology, 80,* 192–201.

CHAPTER 6

Number Flexes for Daily Math Fluency Practice

Number flexes are daily routines that focus on building fluency. They are called number flexes because they build flexibility. These routines support the development of accuracy, flexibility, and efficiency. In this chapter, we will share 10 number flexes that can be easily integrated into the math block. Research states that teachers should do fluency practice every day for at least 10 minutes (Institute of Education Sciences, 2021), and these flexes are one way to achieve this.

DOI: 10.4324/9781032614229-6

NUMBER FLEX 1: SUBITIZING

Students look at groups of things and discuss how the see them.	Students subitize the array. They look at it and might say," I see 8 rows of 10. 8 x 10 is 80. I see 80." Another student might say, "I see 5 columns of 8 which is 40 and then I can double that for 80."	

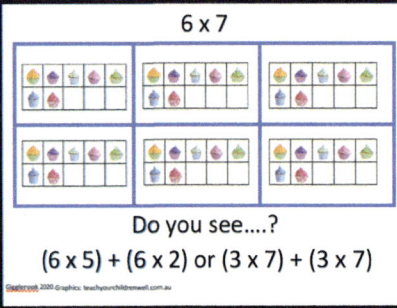

6 x 7

Do you see....?

(6 x 5) + (6 x 2) or (3 x 7) + (3 x 7)

Activity: Subitizing is a routine that is done throughout the grades. The focus of subitizing is for students to recognize quick images at first at a perceptual level (seeing and recognizing small amounts) and then at a conceptual level where they can talk about the ways in which they see the given number such as 2 groups of 5. Clements points out that "this more advanced ability to group and quantify sets quickly in turn supports their development of number sense and arithmetic (1999, p.401)."

For example: Teacher flashes the dice:

Students might say: I see 3 groups of 7. Or, I see 3 groups of 6 and 1 more group of 3.

I can statement:
I can reason about numbers and explain and justify my thinking. I can listen to, understand and respond to the thinking of others and decide whether or not their reasoning makes sense

Purpose
➤ **Reason**
➤ **Generalize math concepts**
➤ **Defend their thinking**

Materials and Tools
Teachers should use both the concrete cards as well as the virtual sets for subitizing.

Concrete Cards	Virtual Sets

Protocol
Overview: The teacher flashes a number card and the students have to state what they saw. The teacher flashes a number on the cards or digitally. The teacher tells the students to take "private think time" to think about what they saw. The teacher asks the students "What did you see?" Notice that this is different than "How many were there?" The first question gets at breaking apart the numbers and analyzing them. Students respond with things like "I see 2 groups of 2".

Questions:

• **What do you see?**
• **How did you see it?**
• **Is there another way to see it?**

FIGURE 6.1 Number Flex 1: Subitizing

NUMBER FLEX 2: WHAT DOESN'T BELONG?

<table>
<tr><td colspan="2" align="center">**What Doesn't Belong?**</td></tr>
<tr>
<td>Activity: The What Doesn't Belong Routine is a routine that focuses on reasoning across different mathematics topics. Students are presented with a variety of options and they have to figure out *which one is not like the others*. They can work by themselves, in pairs, small groups and as a whole group to determine what doesn't belong. The most powerful examples that encourage rich math talk are when there is a reason for each item not to belong.</td>
<td>**I can statement:**
I can reason about numbers and explain and justify my thinking. I can listen to, understand and respond to the thinking of others and decide whether or not their reasoning

Purpose:
➢ **Reason**
➢ **Generalize math concepts**
➢ **Defend their thinking**</td>
</tr>
<tr>
<td>

2 x 5	2 + 2 + 2
3 x 3 + 1	100 ÷ **10**

</td>
<td>**Materials and Tools**
Students should have a variety of tools to think and reason about the ideas being discussed. They should use their toolkits. Students can use manipulatives, drawings and mental math.</td>
</tr>
<tr>
<td>**Protocol**
Overview: The teacher puts the square template on the board and asks the students what doesn't belong. The students have to reason about what doesn't belong and justify their answer. Sometimes there is more than one answer.</td>
<td>**Questions:**

● **How do you know that?**
● **Are you sure about that?**
● **Can you prove it?**
● **Can you show me another way?**
● **Does that make sense?**
● **Does this always work?**</td>
</tr>
</table>

FIGURE 6.2 Number Flex 2: What Doesn't Belong?

NUMBER FLEX 3: SORT THAT

Sort That	
Activity: This routine is about getting students to notice number relationships and provide students the opportunity to choose and use strategies based on the numbers given. The teacher writes several facts and several categories on the board. Students raise their hands and talk about which strategies they would use to solve which facts.	**I can statement:** **I can** think flexibly about numbers. **Purpose** • **Think strategically about numbers** **Materials and Tools** Students should have a variety of tools to think and reason about the numbers being discussed. They should use their toolkits. Students can use manipulatives, drawings and mental math.

2 x 5	12 x 4	3 x 10
3 x 8	4 x 7	8 x 8

Protocol Overview: The teacher writes different equations on the board and then asks the students how they would sort them. Students raise their hand and explain their thinking. Maria: When I see 5 x 9, I think half of 10. So if 10 x 9 is 90 half of it is 45. Carl: Well when I see 9 x 7 I think that is really hard. I would use 10. I know that 10 x 7 is 70 so 1 set less would be 63. Miguel: When I see 8 x 7 I think about my 4's. I use the double double double strategy. I double my 4's which would be 28 plus 28 and that is 56.	**Questions:** ✔ **How could you think about this problem?** ✔ **If you were stuck what could you do?** ✔ **What do you notice about these numbers?**

FIGURE 6.3   Number Flex 3: Sort That

NUMBER FLEX 4: NUMBER TALKS

Number Talks	
Activity: Number talks are teacher led discussions that occur for about 5 -10 minutes where students discuss different ways to solve problems. The teacher gives the students an equation to think about and solve mentally. Then students share and discuss their strategies. $$3 \times 8$$	**I can statement:** *I can multiply and divide numbers using strategies.* **Purpose:** • Think and reason about numbers. • Develop flexibility, efficiency, accuracy and automaticity with whole numbers, fractions and decimals. **Materials and Tools** Students should have a variety of tools to think and reason about the numbers being discussed. They should use their toolkits. Students can use manipulatives, drawings and mental math.
Protocol: The teacher gives the student an expression and the students talk about different ways to solve the problem. For example: **Teacher:** Let's talk about 6 x 7 today. If you have a strategy show me 1 finger, if you have 2 strategies show me 2 fingers. (wait time) Ok. Who has an answer? **Teacher records answers:** 42, 48 (without any reaction to responses) Ok, who wants to defend one of the answers. **Student A:** I did 6 x 6 is 36 and one more 6 is 42. **Student B:** I know that 6 x 5 is 30 and 6 x 2 is 12 and that makes 42. **Student C:** I skip counted up by 6's. **Teacher:** Ok let's think and talk about these strategies. Who wants to describe the strategy they used? **Mica:** I skip counted up.	**Questions:** ✔ **Who wants to defend their thinking?** ✔ **What did you do exactly?** ✔ **Can you lead us through the steps?** ✔ **Tell us why you did that?** ✔ **What are some different ways that you might solve this problem?**

FIGURE 6.4 Number Flex 4: Number Talks

NUMBER FLEX 5: TRUE OR FALSE?

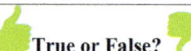
True or False?

Activity: True or False is a routine that focuses on reasoning. Students are presented with mathematical statements that are either true or false. They must reason about those statements, with themselves, in pairs, small groups and as a whole group to determine whether or not the statements make mathematical sense.	**I can statement:** **I can** reason about numbers and explain and justify my thinking. I can listen to, understand and respond to the thinking of others and decide whether or not their reasoning makes sense. **Purpose**

True	False
$3 \times 5 = 15$	$12 = 4 \times 4$
$4 \times 2 = 9 - 1$	

$14 \div 2 = 8$

- Reason alone
- Reason with others and follow their thinking
- To determine whether something is true or false
- To defend one's thinking
- To defend the thinking of another

Materials and Tools
Students should have a variety of tools to think and reason about the numbers being discussed. They should use their toolkits. Students can use manipulatives, drawings and mental math.

Protocol
Overview: The teacher puts some mathematical concept on the board and the students have to vote whether or not it is true or false.

- The teacher puts a mathematical concept on the board.
- The teacher tells the students to take "private think time" to think about the concept.
- After about 30 seconds the teacher tells the students to "turn and talk to a neighbor."
- Everyone comes back together and students raise their hand:
    - thumbs up if they think it is true
    - thumbs down if they think it is false
    - thumbs sideways if they are not for sure.

The teacher calls on various students to explain their thinking.

Questions:

- **How do you know that?**
- **Are you sure about that?**
- **Can you prove it?**
- **Can you show me another way?**
- **Does that make sense?**
- **Does this always work?**

FIGURE 6.5   Number Flex 5: True or False?

NUMBER FLEX 6: NUMBER STRINGS

Number Strings	
10 × 9	
9 × 9	
10 × 8	
9 × 8	
10 × 7	
9 × 7	
Game: Number strings is a routine where the teacher sets out an intentional string of numbers with a set of relationships. Students work their way through the string and then they are given a challenge problem but they have been set up along the string to be able to answer it.	**I can statement:** I can use facts I know to help with facts I don't know. **Purpose:** To work on fluency **Materials and Tools:** Usually students just talk through it without any aids, but thinking udl some students might need to write it out…
Protocol Overview: The students work together as a whole class to discuss the string. The teacher leads the discussion asking students to make connections between what they have just solved and what is on the board currently.	**Questions:** What strategies did you use today? What facts were easy? What facts were challenging?

FIGURE 6.6   Number Flex 6: Number Strings

NUMBER FLEX 7: MATH FLIPS

<table>
<tr>
<td colspan="2" align="center">**Math Flips (Berkley Everett)**
https://vimeo.com/312626727
https://berkeleyeverett.com/math-flips/</td>
</tr>
<tr>
<td>*Activity: Math flips are visual flashcards that help students to see the relationships between facts. There are math flips for all the operations and even fractions. Of course here, we will be focusing on multiplication and division. Everything is available in Google Slides and available to print. They start with problem A and then show problem B and discuss the relationship.*

A

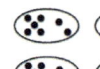   B
</td>
<td>**I can statement:**
I can reason about numbers and explain and justify my thinking. I can listen to, understand and respond to the thinking of others and decide whether or not their reasoning makes sense.
Purpose:
➢ Reason about numbers
➢ Explain their thinking
➢ Justify their answers
Materials and Tools
The slides. Some students might want to have something to write with so this should be an option.</td>
</tr>
<tr>
<td>**Protocol**
Overview: The teacher shows the first problem. Students would comment on how it is 2 groups of 7. Some students might talk about how they see 10 and 4. Then the teacher shows the next slide and students now talk about how they see 4 groups of 7. They discuss how they can use what they know about 2 groups of 7 to help them solve 4 groups of 7.</td>
<td>Questions:
1. What do you notice about this problem. What might the equation be?
2. What do you notice about the 2nd problem. What might the equation be?
3. What is the relationship between the first problem and the second problem.</td>
</tr>
</table>

FIGURE 6.7 Number Flex 7: Math Flips

NUMBER FLEX 8: CUBE CONVERSATIONS

This is work by Steve Wyborney (a math genius). He graciously allowed us to discuss 2 of his number sense routines. Check them all out at http://www.stevewyborney.com/
He also has a free full multiplication course located at:
https://stevewyborney.com/category/multiplication.

Activity: In this routine, students have to discuss different ways to name this number. For example: Student A: I see 3 + 3 + 3 + 1 Student B: I see (3 × 3) + 1 Student C: (2 × 4) + 2 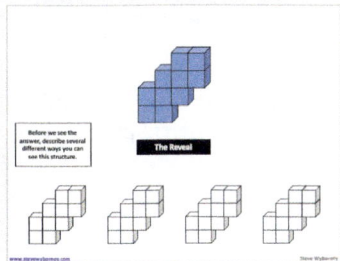	**I can statement:** **I can** reason about numbers and explain and justify my thinking. I can listen to, understand and respond to the thinking of others and decide whether or not their reasoning makes sense. **Purpose:** ➢ Reason ➢ Explain and Justify thinking ➢ Composing and decomposing numbers **Materials:** The Cube Conversation powerpoints. There are several different versions. There are 5 different powerpoints for each one.
Protocol: Teacher shows the powerpoint. Students discuss it. They talk and discuss the different ways that they see it. There are over 80 levels.	**Questions:** 1) What do you see? 2) How do you see it? 3) Is there another way?

FIGURE 6.8 Number Flex 8: Cube Conversations

NUMBER FLEX 9: THE POWER OF COLOR

This is work by Steve Wyborney as well. He graciously allowed us to discuss 2 of his number sense routines. He has so many others. Check them all out at http://www.stevewyborney.com/

Activity: The students get a template Teacher gives the students paper with the circles and crayons. The students shade the circles in different colors and write down equations to represent those shadings. 	**I can statement:** I can model equations. **Purpose:** ➢ Reason ➢ Explain and Justify thinking ➢ Composing and decomposing numbers **Materials:** Splat powerpoints. There are several different versions. There are 5 different powerpoints that
Protocol: The teacher would give the students this grouping of circles. The students would shade the circles. The idea is that students begin to look at groups within groups. We want them to be able to verbalize that. Also, that students can record their thinking in an equation. **5 groups of 2** **2 groups of 5** $5 \times 2 = 10$ $2 \times 5 = 10$ After they shade the circles, they discuss the equations that match their representations.	**Questions:** 1) Tell us what you saw. 2) Did anybody see it a different way? 3) What are some other possibilities? 4) What would be a really long equation that you could make with this? 5) What could be an equation? 6) What could be another equation?

FIGURE 6.9 Number Flex 9: The Power of Color

NUMBER FLEX 10: MOBILES

Mobiles	
Activity: We love this website! https://solveme.edc.org/ *We love that we can see so many different examples and that we can also make our own! We love to get students to make their own on paper and then come up and share them with the class!*	**I can statement:** **I can** reason about numbers and explain and justify my thinking. I can listen to, understand and respond to the thinking of others and decide whether or not their reasoning makes sense.
Protocol Overview: Pull up mobiles that have some of the values filled in and they have to use that information to figure out what the other values are. After you do that you begin to make your own, which you can post, and also the students begin to make their own. For example in this one, we know the total is 18. It is balanced. So both sides must be 9. We see that on the left there are 3 triangles so they must be worth 3 each. That means on right side that the triangle is 3 and the square must be 6. 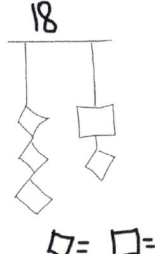	**Purpose:** ➢ Reason ➢ Explain and Justify thinking ➢ Composing and decomposing numbers **Materials and Tools** Students should have a variety of tools to think and reason about the numbers being discussed. They should use their toolkits. Students can use manipulatives, drawings and mental math.

	Questions:
	    1. How did you figure out the puzzle?
	    2. What did you do first?
	    3. How did you make your own puzzle?
	    4. What was easy and what was tricky?

FIGURE 6.10   Number Flex 10: Mobiles

KEY POINTS

Number flexes specifically address computational fluency (strategies, models, procedures, flexibility, and efficiency):

- Subitizing with multiplication
- What doesn't belong?
- Sort the facts
- Number talks/British number talks
- Number strings
- True or false
- Math flips
- Cube conversations
- The power of color
- Mobiles

SUMMARY

Every day for 10 minutes, students should work on fluency. It is the distributed practice across time that allows students to develop a strong sense of number. Playing a variety of energizers and routines where students have to think, reason, listen to others, and justify their answers builds communication skills. Doing a variety of activities where there is more than one answer also builds flexibility and efficiency. These 10 structures that we have discussed can be used in any grade, and they can change throughout the year to expand to the current content of study. Teachers should think about where their students are in the learning trajectory and plan to do these standards-based, academically rigorous, engaging activities throughout the year.

REFLECTION QUESTIONS

1. What routines do you currently use specifically to build fluency?

2. What are 3 different routines that you might try immediately from this chapter?

3. How well do you emphasize representation of the facts with different models?

CALL TO ACTION

 1. Share your favorite "Aha!" moment about teaching multiplication and division number flexes on social media to help spread the movement! #FDJH

 2. Take a photo of different ways that you are teaching multiplication and division number flexes in your classroom and share it on social media to encourage other teachers to do it too! #FDJH

3. Check out these puzzles where students have to add, subtract, multiply, and divide!

REFERENCE

Institute of Education Sciences. (2021). *Assisting students struggling with mathematics: Response to Intervention(RtI) for elementary and middle schools.* Retrieved February 3, 2022, from https://ies.ed.gov/ncee/wwc/practiceguide/2

CHAPTER 7

Assessment

Teachers who use timed tests believe that the tests help children learn basic facts. Children who perform well under time pressure display their skills. Children who have difficulty with skills, or who work more slowly, run the risk of reinforcing wrong learning under pressure. In addition, children can become fearful and negative toward their math learning.

(Burns, 2000, p. 157)

Assessment is a pivotal piece of the instructional cycle. There are many different types of assessment. In this chapter, we want to look at how to develop an assessment cycle that informs what we do and how we do it every day. Students should be part of the process and involved in the conversations and plans based on the assessments. We will use the 5 elements of mathematical proficiency (Kilpatrick et al., 2001) to inform our discussion: conceptual understanding, procedural fluency, adaptive reasoning, strategic competence, and productive disposition. Within this discussion, we will explicitly examine the ways to assess the 4 elements of fluency—accuracy, flexibility, efficiency, and appropriate strategy selection. Through this examination, we will see how students eventually become automatic in their thinking.

"Informative assessment isn't an end in itself, but the beginning of better instruction" (Tomlinson, 2007/2008, p. 11). It should permeate our day-to-day in schools. It should frame our teaching and learning conversations. It is a guide that helps us and our students to know what to do next.

DOI: 10.4324/9781032614229-7

ASSESSING CONCEPTUAL UNDERSTANDING

Assessing for conceptual understanding involves looking at what students know about the concept. Do they understand the math that they are doing? Can they explain the math that they are doing? In other words . . .

Do they KNOW IT? Can they SHOW IT? Can they USE IT?

Here are some examples:

SHORT ESSAYS

There are different types of short essays that you can do with students. You can give them one of these prompts and then they can write about it (see Figure 7.1).

What is multiplication? Show your thinking with numbers, words	What are square number facts?	What are different ways to model multiplication?
What is division? Show your thinking with numbers, words and pictures.	What are different ways to model division?	Why do we use strategies for multiplying and dividing?

FIGURE 7.1   Short Essays

There is also a twist to this structure called the one-minute essay (see Figure 7.2).

Format for 1-Minute Essay

Question example: Describe multiplication with numbers, words and pictures.

Step 1: Give the students a writing prompt.

Step 2: Let them write for 1 minute.

Step 3: They then stop and switch papers with someone.

Step 4: Give the students "1 minute" to write during the switch – they write in a different color and add to their partner's essay.

Step 5: Then "Switch" Back. The students have another 30 seconds to add to their essay.

FIGURE 7.2   One-Minute Essay

ASSESSING PROCEDURAL FLUENCY

Assessing procedural fluency gives the students an opportunity to show that they know *how* to do the math they are doing (see Figure 7.3). These types of assessments require that students do the math, explain what they are doing, look at the math that others are doing and give helpful observations, ask good questions, and give helpful advice on procedures.

What does it mean to use a double-double strategy in multiplication?	What is 9 × 8? What are some different ways to think about it?	Can you explain how to use 10 × 5 to help you with 5 × 5?
Can you explain how knowing your 2's could help with your 4's?	What is the answer to 8 × 7? How did you do it? How do you know that you are correct?	Can you explain how to use the 3's to help with the 6's?

FIGURE 7.3 Assessing Procedural Fluency

When students have a solid understanding of the concept and they have been working with the procedure for a while, we eventually expect them to have "automaticity." Automaticity is the instant recall of a fact. Research defines it as knowing the answer within 3 seconds (Kling & Bay-Williams, 2015). More recently, scholars have argued that it isn't about the 3 seconds but more specifically about if students can answer within a "reasonable amount of time" (Bay-Williams & SanGiovanni, 2021).

Why is instant recall, or automaticity, so important? If students don't have automaticity, especially when working with multi-digit numbers, they get so bogged down calculating each individual fact that their working memory is overloaded, leaving no space to think through the rest of the problem (Bjorklund et al., 1990). When students know their facts automatically, it frees up cognitive space to concentrate on the other parts of a math problem (Poncy et al., 2007). Automaticity (instant recall) is a result of continued, engaging practice over time. It is a result of becoming fluent with your facts. The research resoundingly states that computational fluency is multi-dimensional (accuracy, flexibility, efficiency, and strategy selection) (Kilpatrick et al., 2001; Brownell, 1956/1987; Brownell & Chazal, 1935; Carpenter et al., 1998). As Boaler (2015) points out, "The brain researchers concluded that automaticity should be reached through understanding of numerical relations, achieved through thinking about number strategies" (Delazer et al., 2005).

> We strongly believe that students learn and own their basic facts through a combination of exposure to others, working it out for yourself, playing with concrete materials, experimenting with different forms of representation, and then rehearsing the acquired knowledge unit within your immediate memory, transferring it into long-term memory, and having it validated thousands of times.
>
> (Hattie & Yates, 2014, p. 57)

ASSESSING ADAPTIVE REASONING

Assessing reasoning is about seeing if students can think logically. So, asking them questions that require them to think and explain their reasoning is an important part of the assessment cycle. Many assessments focus heavily on procedural fluency, but it is critical that we value all 5 elements of math proficiency. This means that assessment needs to measure more than just procedural fluency. Worked examples are great ways to engage students in conversations that promote reasoning. Listening to what students share allows us to know and understand how students think and reason.

WORKED EXAMPLES

Worked examples are activities where the students are literally given a sample problem that has been solved. They are told whether the worked example is correct or incorrect and are asked a series of questions about it (see Figures 7.4 and 7.5). Immediately after this, they are given another example to practice. The caveat here is not to give the incorrect worked example to students before they fully understand the actual concept. these are great activities to do with students to get them to reason about the math they are doing. Worked examples should be given and discussed in depth, and then a new problem is given to the students to practice. It is important to center the discussion around the "why" and not to just discuss the "what" of the worked example (McGinn et al., 2015). Through the worked example, the teacher is trying to further deepen conceptual understanding and procedural fluency of the topic through reasoning.

On the Zones Math Blog (n.d.), there are great examples of modifying worked examples for English-language learners. We added the word bank that they have on their examples because we think that is a great idea for all learners. Remember that math is a 2nd language for native English speakers and a 3rd language for students that are learning English. It is important that we scaffold math language at all grades for all students.

Original Problem	Worked Answer	Question	Corrected Answer	New Prompt
$3 \times \underline{\ \ } = 6$	Joe said the answer was 18. He is incorrect.	Who can tell me why he is incorrect? Why do you think he did that? What should he have done?	So Nancy is saying that he multiplying 3×6. That is how he got 18. But 3×18 does not equal 6. So we have to go back to the idea that the equal sign means "the same as". We need to make this side (the left side) the same as the right side. So 3 times what number makes 6? How could we figure that out? How could we use our tools to model our thinking?	Well let's try this one: $4 \times \underline{\ \ } = 8$

FIGURE 7.4 Worked Example 1

What's the Mistake?

Look at this problem.
Kayla made a mistake.
Do you see her mistake?
What did she do wrong?
How can we fix it?

Words to use:

Multiplication

Equal sign   Multiply

Same as   Factor

Product

12

$4 \times$ ___ $= 8$

Explain what she did wrong. Now do it correctly.

$4 \times$ _____ $= 8$

FIGURE 7.5 Worked Example 2

ASSESSING FOR REASONING THROUGH QUESTIONING

There are several types of problems and questions that we can ask our students (see Figures 7.6 and 7.7). Depth of knowledge is a framework that encourages us to ask questions that require that students think, reason, explain, defend, and justify their thinking (Webb, 2002). Here is a snapshot of what that can look like in terms of fluency work.

	What are different strategies and models that we can use to teach arrays?	What are different strategies and models that we can use to teach division?	What are different strategies and models to teach missing numbers?	What are different strategies and models to teach the distributive property?
Dok Level 1 (these are questions where students are required to simply recall/reproduce an answer/do a procedure)	Draw the array.	Draw a picture to model division.	Solve: $3 \times ? = 12$	Split this array: 5×7
Dok Level 2 (these are questions where students have to use information, think about concepts and reason) This is considered a more challenging problem than a level 1 problem.	Draw this array in 2 different ways and explain your thinking.	In this word problem are we looking for how many are in each group or how many groups? How do you know? Explain your thinking.	Solve: $3 \times ? = 12$ Model your thinking and explain your answer	Split this array in 2 different ways and explain your thinking: 5×7
Dok Level 3 (these are questions where students have to reason, plan, explain, justify and defend their thinking)	Draw at least 3 different arrays for 18. Explain your thinking with numbers, words and pictures.	Write and solve a division word problem.	Solve: $3 \times ? = 12$. Tell a word problem for this equation. Model the problem and defend your answer.	Make up your own problem. Split the array in 2 different ways. Explain your thinking. Defend your answer.

FIGURE 7.6 Depth of Knowledge Examples

(Newton, R. (2021). Guided Math Lessons in Third Grade. Getting Started. Routledge. NY.)

Dok 1	Dok 2 **At this level students explain their thinking.**	Dok 3 **At this level students have to justify, defend and prove their thinking with objects, drawings and diagrams.**
What is the answer to ??? Can you model the expression? Can you identify the answer that matches this equation?	How do you know that the equation is correct? Can you pick the correct answer and explain why it is correct? Here is the model, what is the question? What is another way to model that problem? Can you model that on the ....??? Give me an example of a ...type of problem.... Which answer is incorrect? Explain your thinking?	Can you prove that your answer is correct? Prove that... Explain why that is the answer... Show me how to solve that and explain what you are doing. Defend your thinking. Convince me that you are correct.

FIGURE 7.7 Asking Rigorous Questions

(A great resource for asking open questions is Marion Small's good questions: Great ways to differentiate mathematics instruction in the standards-based classroom (2017) The Kentucky Department of Education (2007) has a great document illustrating DOK matrices)

STRATEGIC COMPETENCE

Assessing for strategic competence involves looking at students' thinking about how to work with numbers (see Figure 7.8). We are looking to see if students are flexible and efficient. We are looking to see if they can engage in appropriate strategy selection. Being flexible requires that they can think about numbers in a variety of ways. It means that they can look at structure and pattern and think about different ways to take apart, put together, and combine numbers. Efficiency means that students can see quick and easy ways to take apart, put together, and combine numbers. These two things actually go hand in hand, meaning that in order to be efficient, you have to be flexible. It is flexibility that allows students to be efficient. Here are some prompts to assess for both these things.

Flexibility	Flexibility	Efficiency	Accuracy/ Instant Recall
How can we multiply 8 × 7 in two different ways?	If your friend were stuck on multiplying 5 × 7, what would you tell them to do?	What is a quick way to solve 9 × 4?	What is 2 × 9?

FIGURE 7.8 Four Components of Fluency

Students can take a test and reflect on how they are doing with the different components of fluency (see Figure 7.9).

Fluency Self-Check

Name: _____
Date: _____

After you take the quiz, think about the statements in the table below. Check the boxes to show what you think.

	Doing Great! 	Doing Good! 	Working on it! 
Accuracy I get the right answer often.			
Flexibility I can think about a problem in more than 1 way.			
Efficiency I can decide on quick ways to solve a problem.			
Instant Recall I know the answer right away.			

I feel _____ about learning my math facts.
The easy part is …

The tricky part is …

FIGURE 7.9 Fluency Self-Check

PRODUCTIVE DISPOSITION

A productive disposition is just as important as the other 4 elements. We need students to be confident, competent, risk-taking mathematicians. Students must learn in environments where they are required to be public mathematicians—meaning to think out loud, discuss their ideas, and think about and respond to others' thinking. There are 4 major factors that impact a productive disposition: 1) the teacher, 2) past mathematical experiences, 3) the family, and 4) peers and culture (Mtetwa & Garofalo, 1989).

What happens between the teacher and the students and what happens in terms of the mathematical environment greatly impacts what students can and will do. We have to help students by not "helping them immediately" but by getting them to help themselves. We have to teach them to use their tools to think. We have to let them engage in the productive struggle rather than stealing the struggle. Remember, there are 3 types of struggle: 1) productive struggle, 2) no struggle, and 3) unproductive struggle. We know that when students engage in a productive struggle, they learn and soar to new heights of knowledge. We must scaffold the struggle. In a productive struggle, students grapple with the issues and are able to come up with a solution themselves, developing persistence and resilience in pursuing and attaining the learning goal or understanding, says Jackson. In productive struggles, kids have developed the necessary strategies for working through something difficult. They can also take a teacher's suggestions for help and run with them (Jackson & Lambert, 2010) (http://inservice.ascd.org/how-to-tell-when-learning-struggles-are-productive-or-destructive/).

> Don't Steal the Struggle!
> When students are stuck—don't save them! Let them struggle productively.

We need students to continually think about what they are learning, how they are doing in learning in it, and what they might need to help them. Throughout the learning of facts, students are asked to reflect on what they know, what they are learning, and how they are doing (see Figures 7.10 and 7.11).

What is easy about multiplication?	What is tricky about division?

FIGURE 7.10 Productive Disposition Questions

| Circle the way that you feel when you think about Division? | **Doing Great!** | **Doing Good!** | **Working on it!** |

FIGURE 7.11 Rubric for Division Disposition

Another way we can check on our students' disposition is to imagine the current unit of study as a mountain and have them draw a picture of the mountain and where they currently are on that mountain in terms of the current math topic. One of our friends, Cathy Palkovic, asked her 3rd graders how they felt about division (see Figure 7.12). Here are a few of their responses:

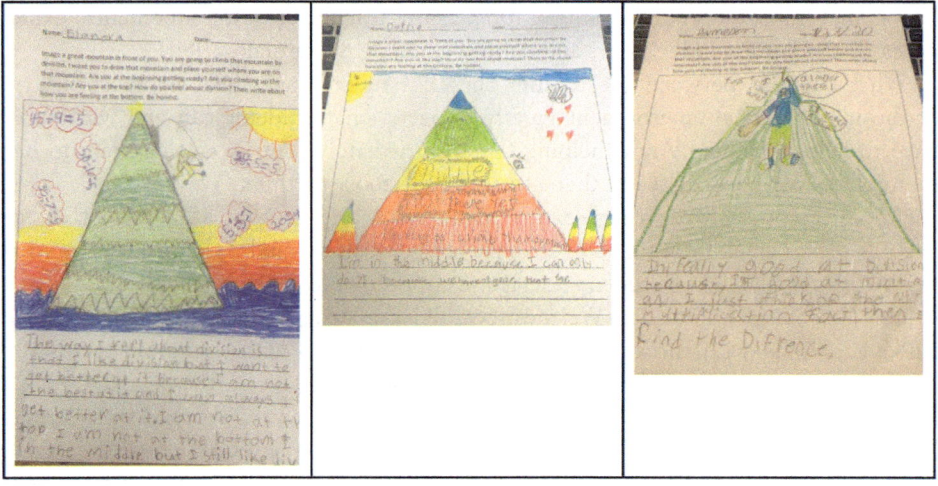

FIGURE 7.12   Division Mountain Pictures

Research shows that students begin to develop a mathematical disposition at an early age. It is therefore essential that we start thinking about and teaching in ways that develop and foster a productive disposition (Ramirez et al., 2013). It is essential that throughout the process of becoming fluent, students are self-monitoring and assessing their behavior. Students should keep track of where they are and where they need to go next (see Figure 7.13).

	Goal	Practicing	Fluent	Notes
COOL!	Multiplying by 0			
WOW!	Multiplying by 1			
GOOD!	Multiplying by 10			
COOL!	Multiplying by 5			
WOW!	Multiplying by 2			
GOOD!	Multiplying by 4			
WOW!	Multiplying by 8			
COOL!	Multiplying by 3			
WOW!	Multiplying by 6			
GOOD!	Multiplying by 9			
WOW!	Multiplying by 7			

_____'s Multiplication Fluency Goal Checklist Part 1

FIGURE 7.13 Self-Monitoring Chart

Many people have used the multiplication/division chart to help students keep track of how they are progressing (see Figure 7.14). There are different ways to do this. Give the students a multiplication chart, and as they learn the facts, they color them in. You want students to make connections between the facts. They should eventually see that 2×2 is 4 and $4 \div 2$ is 2.

X	1	2	3	4	5	6	7	8	9	10	11	12
1	1	2	3	4	5	6	7	8	9	10	11	12
2	2	4	6	8	10	12	14	16	18	20	22	24
3	3	6	9	12	15	18	21	24	27	30	33	36
4	4	8	12	16	20	24	28	32	36	40	44	48
5	5	10	15	20	25	30	35	40	45	50	55	60
6	6	12	18	24	30	36	42	48	54	60	66	72
7	7	14	21	28	35	42	49	56	63	70	77	84
8	8	16	24	32	40	48	56	64	72	80	88	96
9	9	18	27	36	45	54	63	72	81	90	99	108
10	10	20	30	40	50	60	70	80	90	100	110	120
11	11	22	33	44	55	66	77	88	99	110	121	132
12	12	24	36	48	60	72	84	96	108	120	132	144

FIGURE 7.14 Multiplication Chart

Another way to do it is to give the students a blank chart. As they learn the facts, then they fill in the facts (see Figure 7.15).

X	1	2	3	4	5	6	7	8	9	10
1										
2										
3										
4										
5										
6										
7										
8										
9										
10										

FIGURE 7.15 Self-Monitoring Blank Chart

MATH RUNNING RECORDS: INDIVIDUAL PROFILES

Math Running Records are the GPS of math fact fluency (see Figure 7.16) (Newton, 2016). They help us to know where students are, where they should be going next, and how to help get them there. There is a Math Running Record for every operation. The Math Running Record has 3 parts. The first part assesses for automaticity and accuracy. The second part assesses for flexibility and efficiency. The third part taps into the student's mathematical disposition.

Math Running Records are given from the end of kindergarten through middle school. You give them to students until they have learned all of their basic facts. It is an interview assessment that tells us about student thinking, student strategy, and student disposition. It is the beginning of the fluency journey. After the data is collected, it should drive the instruction. Teachers now know what ideas to work on when they do routines in small guided math groups, in math workstations, and in homework. Math Running Records allow us to get very specific about student fact power, and they allow us to scaffold steps to success. NCTM (2000) states that "effective math teaching requires understanding what students know and need to learn and then challenging and supporting them to learn it well."

Student Page

0 x 4	8 x 6
1 x 2	3 x 9
10 x 7	6 x 7
5 x 3	9 x 6
2 x 6	7 x 8
4 x 8	4 x 4

FIGURE 7.16 Math Running Record—Multiplication

Let's take a moment to review the recording of a running record with a 4th grader as well as the analysis of what the interview revealed about the student's thinking (see Figure 7.17).

Multiplying by 6 6 x 7	Multiplying by 9 9 x 6	Multiplying by 7 7 x 8	Multiplying Square #'s 4 x 4
What strategy do you use when you multiply by 6?	What strategy do you use when you multiply by 9?	What strategy do you use when you multiply by 7?	What do you do when you are multiplying a number by itself?
___ use doubles (i.e. 3 x 7 doubled) ___ other ___ can't articulate	___ x10 minus the number being multiplied ___ other ___ can't articulate	___ x5 plus x2 ___ other ___ can't articulate	___ skip count ___ other ___ can't articulate
	If I didn't know 9 x 4, what is a way I could think about and solve this problem?	If I were stuck on 7 x 6, what would you tell me to do?	For example: 5 x 5
For example: 6 x 3 6 x 6 6 x 8			8 x 8 9 x 9
	How would you solve 9 x 7?	How about 7 x 4?	
	How about 9 x 5?	How about 7 x 3?	
Do they know this strategy?	Do they know this strategy?	Do they know this strategy?	Do they know this strategy?
No/Emerging/Yes	No/Emerging/Yes	No/Emerging/Yes	No/Emerging/Yes
M6 Level 0 1 2 3 4M 4	M9 Level 0 1 2 3 4M 4	M7 Level 0 1 2 3 4M 4	MS Level 0 1 2 3 4M 4

Part 3	Question Prompts:
Do you like math? What do you find easy? What do you find tricky? What do you do when you get stuck?	That's interesting/fascinating: tell me what you did. That's interesting/fascinating: tell me how you solved it. That's interesting/fascinating: tell me what you were thinking. How did you solve this problem? Can you tell me more about how you solve these types of problems? What do you mean when you say _____? (i.e. ten friends/neighbor numbers etc.)

General Observations (to be filled out after the interview)

Instructional Response:

Fluency Focus areas (circle all that apply): flexibility efficiency accuracy automaticity

What multiplication strategy should the instruction focus on?

M0 M1 M10 MS M2 M4 M8 M3 M6 M9 M7 MS

For their current instructional level, what is the predominate way in which students are arriving at the answers?

0 1 2 3 4M 4 _____

Overall, what is the way in which students calculated the answers?: 0 1 2 3 4M 4

Comments/Notes about gestures, behaviors, remarks:

FIGURE 7.17 Scored Multiplication Math Running Record

Multiplication Running Record Recording Sheet

Student: Student 2   Teacher: _____   Date: May 2020

Part 1: Initial Observations

Teacher: We are now going to administer Part 1 of the Running Record. I am going to give you a sheet of paper with some problems. I want you to go from the top to the bottom and tell me just the answer. If you get stuck, you can stop and ask for what you need to help you. If you want to pass, you can. We might not do all of the problems. I will be taking notes so that I remember what you did. Let's start.

Part 1	Codes: What do you notice?	Initial Observations of Strategies	Data Code Names
0 x 4 ✓ a 5s (pth)	ca fco skf coh skh sc asc dk	0 1 2 3 4M 4	M0 -multiplying by 0
4 x 1 x 2 (a) 5s pth	ca fco skf coh skh sc asc dk	0 1 2 3 4M 4	M1-multiplying by 1
10 x 7 ✓ (a) 5s pth	ca fco skf coh skh sc asc dk	0 1 2 3 4M 4	M10-multiplying by 10
5 x 3 ✓ (a) 5s pth	ca fco skf coh skh sc asc dk	0 1 2 3 4M 4	M5-multiplying by 5
2 x 6 ✓ (a) 5s pth	ca fco skf coh skh sc asc dk	0 1 2 3 4M 4	M2-multiplying by 2
4 x 8 ✓ a 5s (pth)	ca fco skf coh skh sc asc dk	0 1 2 3 4M 4	M4-multiplying by 4
8 x 6 ✓ a 5s (pth)	ca fco skf coh skh sc asc dk 5 x 8 + 8	0 1 2 (3) 4M 4	M8-multiplying by 8
3 x 9 a 5s pth	ca fco skf coh skh sc asc dk	0 1 2 3 4M 4	M3-multiplying by 3
6 x 7 a 5s pth	ca fco skf coh skh sc asc dk	0 1 2 3 4M 4	M6-multiplying by 6
9 x 6 a 5s pth	ca fco skf coh skh sc asc dk	0 1 2 3 4M 4	M9-multiplying by 9
7 x 8 a 5s pth	ca fco skf coh skh sc asc dk	0 1 2 3 4M 4	M7-multiplying by 7
4 x 4 a 5s pth	ca fco skf coh skh sc asc dk	0 1 2 3 4M 4	MS-multiplying Square Numbers

Codes	Types of Strategies	Strategy Levels
a - automatic 5s - 5 seconds pth - prolonged thinking time	ca – count all with fingers, drawings, manipulatives fco - finger counted on skf - skip counted on fingers coh – counted on in head skh – skip counted in head sc - self- corrected asc – attempted to self-correct dk - didn't know	0 – doesn't know 1 – counting strategies by ones or skip counting using fingers, drawings or manipulatives 2 - mental math/solving in head 3- using known facts and strategies 4M- automatic recall from memory 4 – automatic recall and students have number sense

FIGURE 7.17   (Continued)

Part 2: Flexibility/Efficiency

Teacher: We are now going to administer Part 2 of the Running Record. In this part of the Running Record we are going to talk about what strategies you use when you are solving basic multiplication facts. I am going to tell you a problem and then ask you to tell me how you think about it. I am also going to ask you about some different types of facts. Take your time as you answer and tell me what you are thinking as you see and do the math. I am going to take notes so I can remember everything that happened during this Running Record.

Multiplying by 0 0 x 4	Multiplying by 1 1 x 2	Multiplying by 10 10 x 7	Multiplying by 5 5 x 3
What happens when you multiply by zero?	What happens when you multiply by 1?	What strategy do you use when you multiply by 10?	What strategy do you use when you multiply by 5 ?
___ product is 0 ___ other ___ can't articulate	___ product is the number being multiplied ___ other ___ can't articulate	___ understands place value ___ other ___ can't articulate "add a zero"	___ count by 5's ___ other ___ can't articulate
For example: 0 x 1 5 x 0 0 x 7	For example: 1 x 3 10 x 1 1 x 5	For example: 10 x 8 ✓ a 5 x 10 10 x 3	For example: 5 x 7 ✓ 5s 4 x 5 ✓ a 5 x 9 ✓ a 8x5 + 5
Do they know this strategy?	Do they know this strategy?	Do they know this strategy?	Do they know this strategy?
No/Emerging/Yes M0 Level 0 1 2 3 4M 4	No/Emerging/Yes M1 Level 0 1 2 3 4M 4	No/Emerging/Yes M10 Level 0 1 2 3 4M 4	No/Emerging/(Yes) M5 Level 0 1 2 3 4M (4)
Multiplying by 2 2 x 6	Multiplying by 4 4 x 8	Multiplying by 8 8 x 6	Multiplying by 3 3 x 9
What do you do when you multiply by two?	What strategy do you use when you multiply by four?	What strategy do you use when you multiply by eight?	What strategy do you use when you multiply by 3?
___ double it ___ other ___ can't articulate	___ use doubles (i.e. 2 x 8 doubles) ___ other ___ can't articulate	___ use doubles (i.e. 4 x 6 doubled) ___ other ___ can't articulate	___ x2 plus the number ___ other ___ can't articulate
For example: 2 x 4 ✓ a 2 x 8 ✓ a 2 x 7 ✓ a	For example: 4 x 2 4 x 6 ✓ p th skh 4 x 7 6, 12, 18, 24 4 x 8 skh	If I didn't know 8 x 3 what is a way that I could solve this problem? How about 8 x 5? How about 8 x 9?	For example: 3 x 3 3 x 6 3 x 4
Do they know this strategy?	Do they know this strategy?	Do they know this strategy?	Do they know this strategy?
No/Emerging/(Yes) M2 Level 0 1 2 3 4M (4)	No/Emerging/Yes M4 Level 0 1 2 3 4M 4	No/Emerging/Yes M8 Level 0 1 2 3 4M 4	No/Emerging/Yes M3 Level 0 1 2 3 4M 4

FIGURE 7.17   (Continued)

Multiplying by 6 6 x 7	Multiplying by 9 9 x 6	Multiplying by 7 7 x 8	Multiplying Square #'s 4 x 4
What strategy do you use when you multiply by 6? ___ use doubles (i.e. 3 x 7 doubled) ___other ___can't articulate	What strategy do you use when you multiply by 9? ___ x10 minus the number being multiplied ___other ___can't articulate	What strategy do you use when you multiply by 7? ___ x5 plus x2 ___other ___can't articulate	What do you do when you are multiplying a number by itself? ___skip count ___other ___can't articulate
For example: 6 x 3 6 x 6 6 x 8	If I didn't know 9 x 4, what is a way I could think about and solve this problem? How would you solve 9 x 7? How about 9 x 5?	If I were stuck on 7 x 6, what would you tell me to do? How about 7 x 4? How about 7 x 3?	For example: 5 x 5 8 x 8 9 x 9
Do they know this strategy?	Do they know this strategy?	Do they know this strategy?	Do they know this strategy?
No/Emerging/Yes M6 Level 0 1 2 3 4M 4	No/Emerging/Yes M9 Level 0 1 2 3 4M 4	No/Emerging/Yes M7 Level 0 1 2 3 4M 4	No/Emerging/Yes MS Level 0 1 2 3 4M 4

Part 3

Do you like math? Yes

What do you find easy?

What do you find tricky? 8's

What do you do when you get stuck?

ask for help

Question Prompts:
That's interesting/fascinating: tell me what you did.
That's interesting/fascinating: tell me how you solved it.
That's interesting/fascinating: tell me what you were thinking.
How did you solve this problem?
Can you tell me more about how you solve these types of problems?
What do you mean when you say _____ ? (i.e. ten friends/neighbor numbers etc.)

General Observations (to be filled out after the interview)
Instructional Response:
Fluency Focus areas (circle all that apply): flexibility efficiency accuracy automaticity
What multiplication strategy should the instruction focus on?
M0 M1 M10 M5 M2 (M4) M8 M3 M6 M9 M7 MS

For their current instructional level, what is the predominant way in which students are arriving at the answers?
0 1 (2) 3 4M 4 _____ SKh _____

Overall, what is the way in which students calculated the answers?: 0 1 2 3 4M 4
Comments/Notes about gestures, behaviors, remarks:

Updated version of Dr. Nicki Newton's Math Running Records protocol, from *Math Running Records in Action* (Routledge 2016/2019). Teachers have permission to use this with their classroom and the students they work with. Schools, districts and universities have permission to copy these for professional development work. If you are doing other types of training, you must get permission from Dr. Nicki and Routledge (drnicki7@gmail.com).

FIGURE 7.17   (Continued)

In part 1, we are getting a sense of the accuracy and relative speed of the students. As you can see, other than the 1 ´ 2, which was a blip, his accuracy is right on. For the relative speed, this student was thrown a little bit with the ×0 fact, but then was automatic with the ×10, ×5, and ×2. At the ´4 and ×8, though, he showed prolonged thinking. In part 2, we will delve a bit deeper into these strategy zones to learn about what his thinking is.

In part 2, we want to provide additional expressions within each strategy zone to be sure that the student truly has mastery with a variety of examples in each strategy zone.

This is also a time in the interview when we are able to see if the students can articulate their thinking. For the ×10, we want to be sure we don't allow students to say that they "added a zero" when they were multiplying by 10 since mathematically, that's not what is happening, and that rule will expire once they begin multiplying by decimals. Instead, typically in grade 3, we can encourage students to describe 8 × 10 = 80 because it is the same as eight 10's. In grades 4 and up, though, discussing the shifting place value position will help them with multiplying whole numbers not only by 10 but by decimals in grade 5 and up as well.

This student's ×5 facts are in place, and he even described the derived multiplicative strategy of using a nearby fact of 5 × 10 = 50 to help determine 9 × 5, which is so much more efficient than many students use of the skip counting by 5's starting at 5. All his doubles facts are in place as well. It's when he is asked the ×4 facts that he falls back into the skip counting in his head. At this point, Ann Elise stopped this section of the Math Running Record because she knows that she wants this student to begin working on his ×4 facts but explore the relationship of ×4 as double the ×2 with a variety of concrete, pictorial, and abstract representations and then practice them using games.

Part 3 is the last part of the Math Running Record and is a chance for us to talk to our students about their math dispositions. We begin by asking them if they like math. Next, we refer back to the Part 1 benchmark expressions and ask the students which ones look easy to them and which ones look hard. The ones that look hard are because they are unfamiliar to the students, and they don't yet have a way to begin to figure it out efficiently. By learning the facts that look tricky to our students, we can be sure to explore strategies and experiences for those facts that build their confidence. Once they have developed the familiarity with them and have strategies to figure those facts out, then those will no longer look tricky to them.

CLASS LISTS

It is important to know where the students are as a class so that you can form small guided math groups and offer purposeful, differentiated workstations that meet the needs of your students (see Figures 7.18 and 7.19). There are several ways to do this. Dodge (n.d.) discusses using a "system of check-minus, check and check plus or the numbers 4, 3, 2, and 1 to indicate student proficiency with the skill."

Addition Facts Students	Multiplying by 0	Multiplying by 1	Multiplying by 10	Multiplying by 5	Multiplying by 2
Maria A.	x	x	x	x	x
Tyrone B.	x	x		x	
Carol B.	x	x	x		
Nina C.	x	x			
Luke D.	x	x			
Abdu D.	x	x	x	x	

FIGURE 7.18 Sample Classroom Progress Chart

From this list, we can see what students know and what they need to learn. We can also decide on the groups that need to be pulled and what they need to work. We can see that Tommy, Nina, and Luke should all be working on multiplying by 10. We can see that Carol, Nina, and Luke need to also work on multiplying by 5. We can see that most of these students are struggling with multiplying by 2.

	A			O		P	
	Student (choose from lis			Current Stra		Current Leve	
	Bailey, David	▾		02 - M1	▾	0	▾
	Moore, Karen	▾		02 - M1	▾	0	▾
	Nichols, Michael	▾		03 - M10	▾	1	▾
	Styles, Susanne	▾		04 - M5	▾	1	▾
	Duquette, Angela	▾		04 - M5	▾	1	▾
	Frizzell, Emily	▾		04 - M5	▾	1	▾
	Griffin, Sue	▾		04 - M5	▾	2	▾
	Holt, Heidi	▾		04 - M5	▾	2	▾
	Manning, Benjamin	▾		06 - M4	▾	1	▾
	Rancourt, Rosemary	▾		06 - M4	▾	1	▾
	Record, Daniel	▾		06 - M4	▾	1	▾
	Rodrigues, Justina	▾		06 - M4	▾	1	▾
	Schersten, Tom	▾		07 - M8	▾	1	▾
	Vig, Elizabeth	▾		07 - M8	▾	2	▾
	Williams, Kylynn	▾		07 - M8	▾	2	▾
	King, Julie	▾		07 - M8	▾	3	▾
	Chappell, Cordelia	▾		07 - M8	▾	3	▾

FIGURE 7.19 Math Running Record Class Strategy Summary Chart

Once we have administered Math Running Records to all our students, we can then develop our plan for the instructional response so that all students are having purposeful practice in their current strategy zone. In Figure 7.18, you can see a snapshot from an example classroom using a data file available to you at https://tinyurl.com/MathRR. We can see that David and Karen need to explore what multiplication means conceptually. Michael is currently skip counting his ×10 facts, so he will need to explore the patterns of the products when we are multiplying by 10. There are then a handful of students who are currently skip counting by 5. Those who have a current level of 1 are skip counting using their fingers, and those who are at level 2 are skip counting mentally. For these students, we will want to explore how ×5 is half of ×10 to determine these facts or to use a ×5 fact you know for sure and use the distributive property to arrive at the product.

As we move into the ×4 facts, the 4 students who will be working on these are all skip counting using their fingers. For these students, we will explore the relationship between ×2 and ×4 facts using concrete, pictorial, and abstract representations. For the 5 students who will be working on their ×8 facts, some are skip counting with fingers or in their heads and will need some work on developing derived multiplicative strategies, but a couple of them are at level 3 and already using derived facts, so they may need to improve their accuracy by practicing with games.

KEY POINTS

- Conceptual assessments
- Procedural assessments
- Strategic assessments
- Reasoning assessments
- Productive disposition assessments

SUMMARY

Assessment should be an ongoing part of the instructional cycle. We need to consciously *assess for learning* so that we know what to do next (Stiggins & Guskey, 2007). Guskey (2007) notes that we must "use assessments as sources of information for both students and teachers" (pg. 16), that our assessments must be followed up "with high-quality corrective instruction," and that we must allow our students the opportunity to practice it until they can "demonstrate success."

We need to make sure that we frame our assessments around the elements of mathematical proficiency. We need to check for conceptual understanding, procedural fluency, strategic competence, adaptive reasoning, and a productive disposition. Our assessment should encourage and motivate students to keep working toward specific goals. Assessment doesn't just happen; we have to plan for it.

REFLECTION QUESTIONS

1. What stands out for you in this chapter?
2. What are your strengths with assessment, and what are some challenges?
3. Can you state or draw a three-step action plan based on this chapter?

CALL TO ACTION

 1. Share your favorite "Aha!" moment about assessing multiplication and division on social media to help spread the movement! #FDJH

 2. Take a photo of different ways that you are assessing multiplication and division in your classroom and share it on social media to encourage other teachers to do it too! #FDJH

3. Download this letter that will give you plenty of resources to get started with Math Running Records!

REFERENCES

Adapted from Kaplinsky. https://robertkaplinsky.com/depth-knowledge-matrix-elementary-math/.

Bay-Williams, J., & SanGiovanni, S. (2021). *Figuring out fluency in mathematics teaching and learning.* Corwin.

Bjorklund, D. F., Muir-Broaddus, J. E., & Schneider, W. (1990). The role of knowledge in the development of strategies. In D. F. Bjorklund (Ed.), *Children's strategies: Contemporary views of cognitive development.* Erlbaum.

Boaler, J. (2015). *Fluency without fear.* Retrieved May 15, 2019 from www.youcubed.org/evidence/fluency-without-fear/

Brownell, W. A. (1956/1987). Meaning and skill: Maintaining the balance. *Arithmetic Teacher, 34*(8), 18–25.

Brownell, W. A., & Chazal, C. B. (1935). The effects of premature drill in third-grade arithmetic. *The Journal of Educational Research, 29*(1), 17–28.

Burns, M. (2000). *About teaching mathematics: A K-8 resource.* Math Solutions Publications.

Carpenter, T. P., Franke, M. L., Jacobs, V. R., Fennema, E., & Empson, S. B. (1998). A longitudinal study of invention and understanding in children's multidigit addition and subtraction. *Journal for Research in Mathematics Education, 29*(1), 3–20. https://doi.org/10.2307/749715

Delazer, M., Isachebeck, A., Domahs, F., Zamarian, L., Koppelstaetter, F., Siedentopf, C. M., Kaufmann, L., Benke, T., & Felber, S. (2005). Learning by strategies an learning by drill-evidence from an fMRI study. *NeuroImage, 25,* 838–849.

Dodge, J. (n.d.). *What are Formative assessments and why should we use them?* www.scholastic.com/teachers/articles/teaching-content/what-are-formative-assessments-and-why-should-we-use-them/

Guskey, T. (2007). Using assessments to improve teaching and learning. In D. Reeves (Ed.), *Ahead of the curve: The power of assessment to transform teaching and learning.* Solution Tree.

Hattie, J., & Yates, G. C. R. (2014). *Visible learning and the science of how we learn.* Routledge.

Jackson, R., & Lambert, C. (2010). *How to support struggling students.* ASCD.

Kentucky Department of Education. (2007). *Support materials for core content for assessment version 4.1 mathematics.* Retrieved January 15, 2017, from https://www.education.ky.gov/AA/Pages/default.aspx

Kilpatrick, J., Swafford, J., & Findell, B. (2001). *Adding it up: Helping children learn mathematics.* National Academy Press.

Kling, G., & Bay-Williams, J. M. (2015). Three steps to mastering multiplication facts. *Teaching Children Mathematics, 21*(9), 548.

McGinn, K., Lange, M., & Booth, L. (2015). A worked example for creating worked examples. *Mathematics Teaching in the Middle School, 21,* 26–33.

Mtetwa, D., & Garofalo, J. (1989). Beliefs about mathematics: An overlooked aspect of student difficulties. *Academic Therapy, 24*(5), 611–618.

National Council of Teachers of Mathematics. (2000). *Principles and standards for school mathematics.* National Council of Teachers of Mathematics.

Newton, R. (2016). *Math running records.* Routledge.

Poncy, B. C., Skinner, C. H., & Jaspers, K. E. (2007). Evaluating and comparing interventions designed to enhance math fact accuracy and fluency: Cover, copy, and compare versus taped problems. *Journal of Behavioral Education, 16,* 27–37.

Ramirez, G., Gunderson, E., Levine, S., & Beilock, S. (2013). Math anxiety, working memory, and math achievement in early elementary school. *Journal of Cognition and Development, 14*(2), 187–202.

Small, M. (2017). *Good questions: Great ways to differentiate math in the standards based classroom.* Teachers College Press.

Stiggins, R., & Guskey, T. (2007). Assessment for learning: An essential foundation of productive instruction. In D. Reeves (Ed.), *Ahead of the curve: The power of assessment to transform teaching and learning.* Solution Tree.

Tomlinson, C. (2007). *Learning to love assessment.* Retrieved December 29, 2021, from https://docs. google.com/document/d/1mMxFcQpBqOsANcDp_5j2rXfnYydWfhSX/edit#

Webb, N. (2002). An analysis of the alignment between mathematics standards and assessments for three states. Paper presented at *the annual meeting of the American educational Research Association.* New Orleans, LA.

CHAPTER 8

Parental Involvement

MAKING THE MOST OF THE HOME–SCHOOL CONNECTION

By the time students reach the end of second grade where the foundations of multiplicative reasoning are introduced, parents have a couple of years of "school math" under their belts. Before you embark on your fluency journey with them, consider where they've been and how their experiences in earlier grades (or even their own experiences as learners themselves) may shape their perspectives regarding mastering the basic facts. In most cases, parents really want to support their kids at home, but research shows that parental involvement in math may be less common than with other subjects such as reading (Sheldon et al., 2010). This is especially true now with all of the changes to the standards. Parents no longer recognize the math that their students are learning. This can make them feel powerless and helpless, which leads to great frustration when trying to help with homework. There are words they don't know, models they don't understand, and typically an exasperated child stating, "That's not how my teacher does it!" Because so many math concepts may be foreign to parents, and we may not expect them to be able to support their children at home, it might seem like a no-brainer that we can at least have them support fluency with basic facts. Those are straightforward, right? Wrong . . . well, sort of. While *parents* may feel confident that this is an easy area to offer support, it's important to know what their "support" looks like and if it complements or contradicts what you are working on in your classroom. Even in this area, we are (or should be) using models, language, and strategies that are unfamiliar to families, and failure to recognize this can make our job harder and result in frustration for students.

While it may be tempting to put out a call to parents to support math facts at home, the fact is that while *we* now know that drill and kill will not get the job done, parents may not. Think about how they probably learned, and what their perception of fluency might be. These factors make it highly probable that if left to their own devices, parents will eagerly break out the trusty flashcards and get their drill on. Worse, they will likely focus on speed, which, again, is what they probably remember as the focus of their own experience. When they time their kids and drill them, it is all with the best intentions, and they may think their efforts are working if they presume that rote recall is what we are going for and their kids should be able to regurgitate answers. We, however, know that is not the goal. Therefore, if we want parental support, we must educate them on what fluency means, share our strategies, and equip them with resources that will support and reinforce what is happening in the classroom. Research indicates that this is an important component of making home support more effective (Van Voorhis, 2007).

DOI: 10.4324/9781032614229-8

GETTING PARENTS ON BOARD

Many parents may need to know the rationale for all of these strategies. Without context, it likely seems progressive, bizarre, and flat out unnecessarily complex. Why not just memorize? In our experience, we are frequently asked why we are taking this approach, since the "old way" worked just fine back in the day when parents were in school. Believe it or not, we welcome that question because if it goes unanswered, it's likely that the support coming from home will undermine what we are trying to accomplish in the classroom. When confronted with that question, or perhaps at open house if you choose to put it out there, you can try this two-part response.

Part One

Give the example from the introduction of the book about "knowing 7×8 but not 7×9" because it makes a strong statement. We certainly never want a student or parent to think that the "times tables" are something to learn without a purpose. That purpose is to be able to attack more complex calculations, discover the relationships between numbers, and learn how to manipulate numbers in different situations. Remember, one of the big goals is flexibility. Parents need to understand this.

Part Two

Ask parents how *they* feel about math. Ask what feelings come to mind when asked to recall their own experiences with math and you will likely hear the groans and complaints. When we have a large audience of parents and caregivers and ask those who love math to raise their hands, it is always the same: a few sad little hands go up. This speaks volumes about the impact of the former standards and the traditional instruction that accompanied them. The exercise is great because it always provides the opportunity to share with parents that we are not here to repeat the mistakes of the past. We want better for their kids and so should they. Try this with a group of parents, or maybe at a party with your friends, and see what happens. In our experience, it always has tremendous impact and ends the debate about why we are changing things up.

SHARING THE STRATEGIES

Research consistently indicates that the beliefs and expectations of parents in math predict student achievement in elementary and middle school (Entwisle & Alexander, 1996; Gill & Reynolds, 1999; Halle et al., 1997; Holloway, 1986). Parents are strangers to these strategies, just as most of us were until recently. If we don't clue them in to what the strategies are, they can't participate in the conversation. They need to know the language. If you have the ability to begin a movement in your school, then we suggest starting right out of the gate in kindergarten. The earlier you get parents on board, the easier it is to generate and sustain the momentum needed to get kids fluent by the end of elementary school.

*Sidebar: If you have reluctant colleagues, go back to the two-part exercise above that you did with parents, and follow it up with questions about how a lack of fluency with basic facts impacts student progress across other areas of mathematics. This will likely get their attention and allow you to make a case for a school-wide effort.

Step One: Share the Following With Parents up Front

The names of the strategies (see Figures 8.1 and 8.2): This is actually an important step, as students refer to strategies by name and parents need to know what their kids are talking about. We are pretty sure that "double double" or "double double double" doesn't come up in everyday adult conversation, and sounds pretty ridiculous, unless you're a teacher!

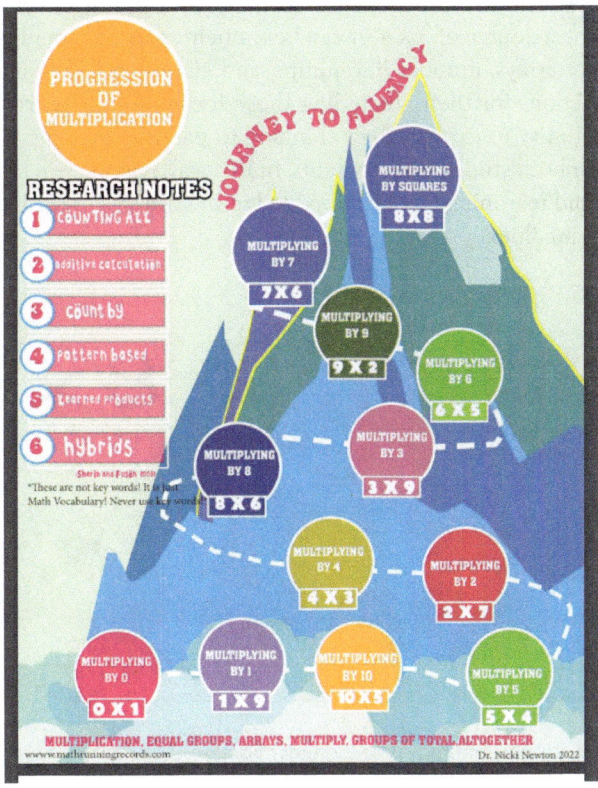

FIGURE 8.1 Multiplication Strategies

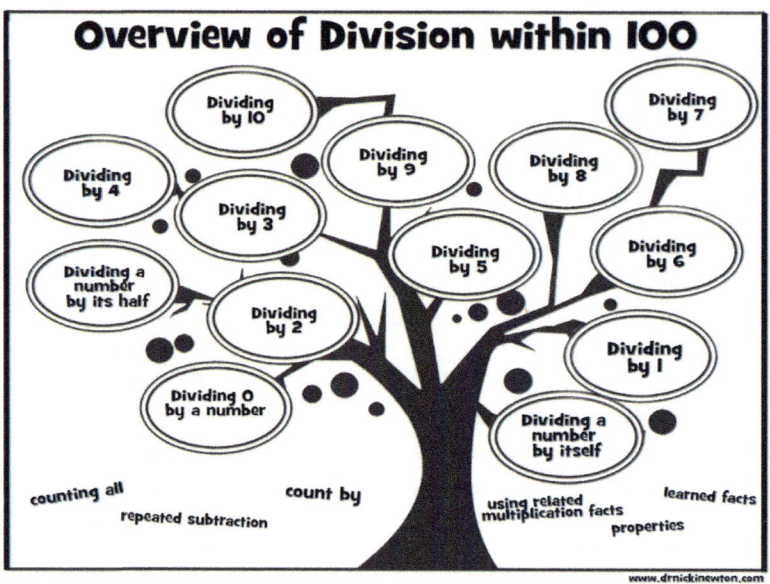

FIGURE 8.2 Division Strategies

What the strategies look like, including examples of the strategies in action: For example, for some facts, more than one strategy may apply. Parents need to recognize that 6 × 4 can be a double of 3 × 4 or can be thought of as a 5 fact plus one more group. Visuals like arrays, number line jumps, and circles and stars are helpful to make the connections for them. Including those for familial support also give students the opportunity to explain their thinking to parents, with the visuals serving as a scaffold for both students and parents. In this way, parents will begin to understand the logic and reasoning that is embedded in the strategies and appreciate their role in developing fluency.

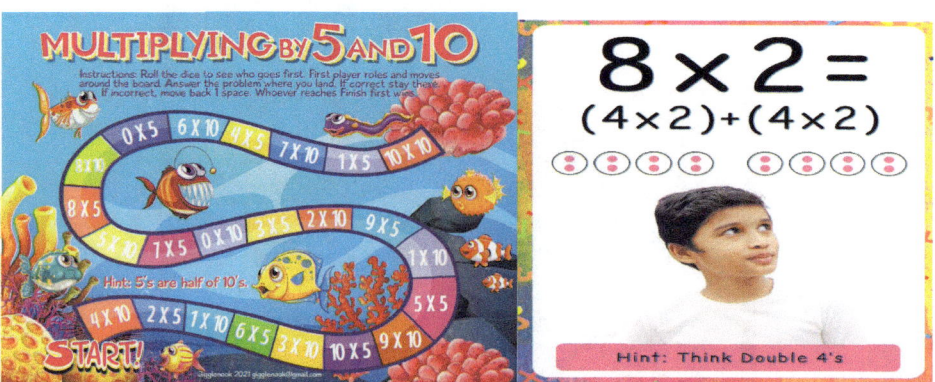

FIGURE 8.3 AND 8.4 Using Strategies to Connect Facts

What strategy instruction looks like in your classroom: What will your students see and hear? What will you use in your instruction? Parents need to know what a Rekenrek is, what an array is, how a number line is used, and whatever else you may be doing. One way of sharing this is to have the students write a weekly memo home explaining what they learned. This may include drawings, key vocabulary, and even photos if students are using the computer. If you want to capitalize on the media literacy of your students, have them produce videos or slides where they explain what they've learned and then have them share with their families. Many schools have platforms that they use to foster two-way communication, which would be great for this. Students truly enjoy it, and it really brings parents and caregivers into the learning process in a non-threatening way.

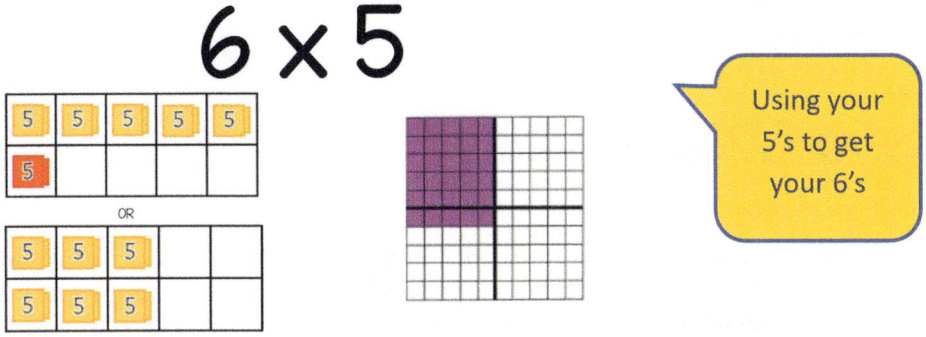

FIGURE 8.5 *Example of a Strategy*

To genuinely help parents grasp the value and importance of fluency with basic facts, it is best to keep the dialogue going throughout the year as strategies are introduced. Continue to provide visuals, specific language, models, and explanations. This could take place in the form of a newsletter, handbook, web page, or series of videos that you produce with your students (see Figure 8.6). (That last suggestion may sound intimidating but can be easily done on a cell phone . . . no need to be fancy!!) If you really want to generate momentum, consider hosting a parent night where you personally explain the progression of facts, take parents through the strategies, and have their kids show them the models used in the classroom. Again, research has indicated that parents who participate in teacher-led training have students who make greater gains in mathematics (Starkey & Klein, 2000; Shaver & Walls, 1998).

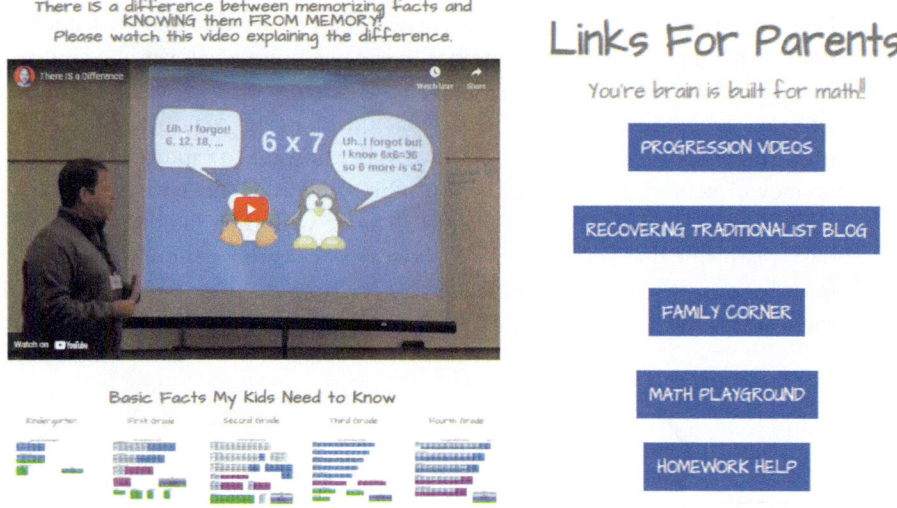

FIGURE 8.6   Parent Page on District Website

(Credit: Foxborough Public Schools)

Recently, a colleague shared a new spin on this called Math Mornings. She didn't use it for fact fluency per se, but you easily could. The idea is to invite parents in at the beginning of the school day (some are already there dropping of their kiddos, so that helps) and give them an overview of everything we have been discussing (strategies, language, models, etc.). You have them as a captive audience for 15 minutes before . . . and this is the brilliant part . . . the students are invited to join you to "teach" their parents. Following the "lesson," they play a game together to practice the basic facts. I have seen these mornings in action, and they are wildly successful, as parents see the flexibility, reasoning, and sense making that happens, not to mention that math feels fun, which is often very different from how they remember learning. If you plan to try this, here are a few basic tips:

- Spend a few days prepping students. They should practice how to "teach" the lesson and know how to play the game.
- Take advantage of student excitement to boost participation rates. When your child is making it clear that they want you to come to the school and play with them, you try your best to make it happen. We have been stunned to see more parents come for these morning events than those we offer in the evening.
- Provide plenty of notice and allow siblings to join. Many parents work or have to care for younger siblings during the day. Letting them know well in advance will increase the odds that they will attend.
- Allow space for adults other than parents to attend. Perhaps a grandparent, cousin, aunt, or uncle is the adult who supports the student at home. Making it clear that the invitation is open to them is important.

- Have a backup plan for students who may not have anyone in attendance. Think about partnering with other teachers in your grade level so you have other adults available, or talk with your principal about freeing up some educational assistants or other adults who can join. The best part is that they will also get the "lesson!"

FIGURE 8.7 A Glimpse at Math Morning in Action

While it will take a bit of time to prepare and execute, you will get a huge return on your investment and get more parents on your team.

Step Two: Create a Homework Plan

One of the best ways to capitalize on the support of parents is to get them involved in the process every night. As noted by Epstein (2001), communicating with parents, getting them involved, and equipping them with the tools to effectively support learning at home increases student success. Letting parents know that practice with basic facts can be fun yet meaningful and purposeful will allow them to see that we have moved beyond the often painful exercises of flashcards or practice tests that it typically has been in the past.

Herein is a menu of possibilities. What works for others may not work for you, so it may take a bit of trial and error and possibly customizing an idea to make it your own before you find exactly what system you will use. You may even want to alternate and employ different ideas at different times to keep things fresh. Keep in mind that successful plans include two-way communication and accountability so that both you and your parents feel valued in the process.

Family games of the week/month: There are many, many games available for free on the internet that you can easily reproduce. They are fabulous for reinforcing the strategy you are working on while at the same time providing important, purposeful practice for your students. Usually all you need are some dice, perhaps a paper clip for a spinner, something to use as a game piece, and you're good to go! This is a favorite suggestion for fact practice because games are engaging, fun, and really get parents involved in the process. Grab some gallon zipper baggies, fire up your laminator, and refer to the collection of examples to get you started! Don't forget the accountability sheets, clear, explicit directions, the math goal ("I Am Learning to" or "I Can" statement), and expectations for frequency of play.

FIGURE 8.8 Samples of Board Games for Home

Strategy cards: A step beyond traditional flashcards, strategy cards reinforce the strategy along with a visual representation (such as a dice model, array, or number line jumps). With strategy cards, you can create rings, wallets, or just use old-fashioned baggies. When creating rings of cards for your students, you can differentiate so that they are only practicing what they have yet to learn, or strategically target a specific progression such as 2's, 4's, 8's. As they master facts, they come off of the ring. The wallets are similar in that the facts they know are on one side and the facts they are still working on are on the other side. It is important that students understand how this works and why certain facts are grouped together so that they can share this intentionality with parents. Making this type of practice intentional is very important. It should not be students just grabbing an entire set of cards and going through them, unless they have mastered all of them and are truly just building automaticity.

If you're about keeping it super simple, baggies labeled "all set" and "not yet" help students organize their cards. If you want to be fancier, you can make a "math fact wallet" from a two-pocket folder. Because they are strategy cards, each time students practice, they reinforce the visual model, the strategy used, and the solution. Whichever option you choose, you should include an accountability sheet to track and monitor practice each night. This could even include a reflection regarding which facts are sticky (as in they are getting them) and which are tricky. This helps students set their own goals and understand where they should focus their attention, while offering you some valuable insight into where they may need support.

FIGURE 8.9   Strategy Cards and Math Fact Wallet

Partner games: With nothing more than a deck of cards, tons of opportunities abound. Whether it be playing War (for addition, subtraction, or multiplication), Salute, Partner Towers, or Fact Family Scatter, kids will love practicing their facts at home (see Figure 8.9). For a really easy partner activity that takes the focus off of answer-getting and onto strategic reasoning, parents can use traditional flash cards or strategy cards but rather than seek the solution, instead question which strategy the child would use and why. Students really seem to enjoy this challenge and it helps to reinforce our approach with families.

FIGURE 8.9 Samples of Salute and Partner Towers games

Get creative: Have students and parents work together to make their own activities (see Figure 8.10). Some examples include making concentration cards (math fact on one card, solution on the other), making board games, or writing questions for a "Who Am I?" game (give clues about a specific fact or strategy). For those interested in movement, suggest passing a ball around like hot potato, but attaching a math fact to it, adapting a game of HORSE at the basketball hoop, or perhaps making up their own kinesthetic game.

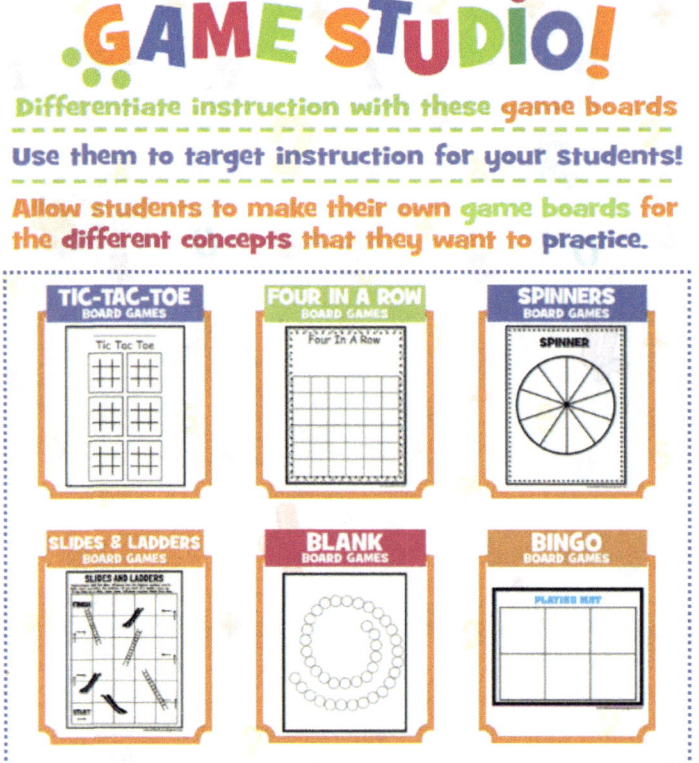

FIGURE 8.10   Sample Make Your Own Game Boards

(Credit: mathfactfluencyplayground.com)

Embrace technology: We are constantly in competition with technology for the attention of our students. If you can't beat them, join them!! Many parents love to pass off their devices to their kids to keep them busy and entertained. By sharing carefully selected apps (way too many are just drill and kill) and/or websites, you will equip parents with the proper tools to work on math facts. This option is perfect for families who may not have the time for some of the others that require direct parental involvement. Many of these applications are adaptive, meaning they meet students where they are after a quick assessment, and many also provide detailed reports of which facts students have mastered and which they need to continue to work on. If you look into some of the sites you use at school, sometimes there is a free version available for parents to use at home. This is great because students get personalized practice using a medium that you have vetted.

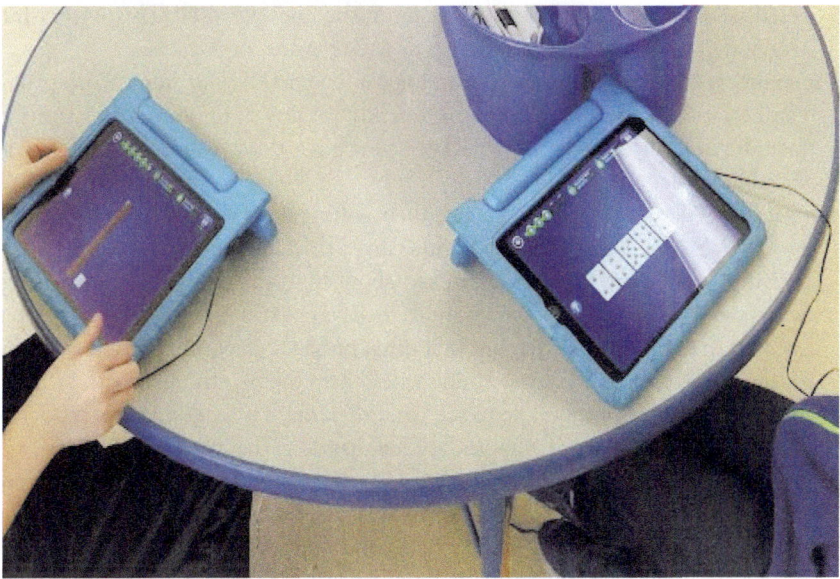

FIGURE 8.11   Technology-Enhanced Fact Practice

PROVIDING ONGOING FEEDBACK

For a successful home–school partnership, it is critical to maintain a dialogue with parents (Epstein, 2001). You are continually gathering formative assessment data regarding the progress of their children in terms of fluency with basic facts. Don't wait until report card time to share what you know! Regardless of how you collect your data (hopefully not timed tests, but no judgements!), if parents know what their kids need to work on, they won't waste time working on what they already know. Keeping them in the loop need not be a laborious process. Here are some suggestions for communicating progress:

Weekly progress note—with a twist: This note home is not written by you but by the student! For younger students, you can create a form letter that they simply fill in. The note can share the strategy they are working on, what facts they have mastered, and which they should focus on in the coming week. This is very informative and very validating for students as they master more facts.

Communication log: A communication log can be part of your homework protocol. It can be a two-way form where you quickly jot down a comment or simply circle facts to help direct parents to where you have noticed growth and where they should practice more.

Pen pal letter: Allow students to practice their writing skills while sharing their reflections on the week. The letter to their parent(s) should include thoughts about the homework activities for the week (what they liked about the games, how they are feeling, what was tricky, etc.), what facts they want to focus on next week, and where they have felt successful. An additional element that often provides great information is a suggestion by the student about what they think would help them with their math facts.

Work samples: Parents always like to see what their child is working on in school. Simply sending home evidence of what the student has done that week can really help parents to see where their efforts are paying off and where they should be spending more time.

Technology: Platforms such as Class Dojo and SeeSaw are great for maintaining communication with parents. Students can also use apps such as Explain Everything to talk through strategies and show how they use tools to support their reasoning. As mentioned earlier, you can also have students produce videos or slideshows outlining their facts and strategies where they can create visuals, add voiceovers explaining their thinking, or incorporate photos.

How does Seesaw work?

Teachers find or create activities to share with students.

Students take pics, draw, record videos and more to capture learning in a portfolio.

Families see their child's work and leave comments and encouragement.

FIGURE 8.12 SeeSaw Platform for Parents

ADDRESSING PARENTAL CONCERNS

With clear grade-level objectives in most states regarding fact fluency, it is normal for parents to grow concerned when it appears that their child not progressing at the anticipated rate. This is definitely worse in schools where standards-based report cards are not in place and grades seem to carry a heavy weight. As students transition from additive to multiplicative reasoning, this is even more pronounced, and this is not a simple shift (Malola et al., 2020). If students have yet to master their addition facts, should they be moving to multiplication? Will they be behind if they don't begin with their peers in 3rd grade?

When parents get panicky about this and want to provide a heavy dose of intervention, it can be a slippery slope. In their quest for perceived proficiency, parents may start to drill the flashcards. They may not recognize or honor the time it may take for students to engage in a strategy before producing the solution and get frustrated because an answer is not produced instantaneously. This will also frustrate the student and could make them feel unsuccessful. Although it is the goal for students to become automatic, we know that there are stages that students progress through, and that they will not all progress at the same rate (Hurst, 2015; Baroody, 2006). While there may be different levels of intervention and support available in your school, there are certain ideas that you can share with parents to put their minds at ease.

First and foremost, parents need to understand that children develop at different rates and that each brain is unique. Did the student arrive in your classroom with the expected prerequisite skills? Have they mastered their basic addition and subtraction facts? If not, that's a wonderful place to begin and helps parents to have a targeted focus. Reassure parents that this is okay, but that it is important to have mastery, as the work with multiplication and division builds upon and connects with those addition and subtraction facts. Again, take the opportunity to reassure them and remind them that students have strategies for addition and subtraction and that those should be part of the process.

If the prerequisite skills are solid, think about different strategies for practice that cater to the learning style of the student. This may include more hands-on manipulation of objects. Parents should know that you don't need specific "math objects" such as snap cubes or two-sided counters to engage in this type of practice. Point out that objects around the house (such as pasta noodles, candy, beans, tiny toys, etc.) are perfect and probably more engaging!

If their child would benefit from additional written practice, share some ideas with parents so that it doesn't become a rote exercise. Perhaps they can use shaving cream, write on a tablet, or use sidewalk chalk. Written practice does not have to mean worksheets or white lined paper! Coaching parents on what this could look like will help to ensure that practice doesn't feel like punishment for students *or parents*!

If a student would benefit from daily review, suggest making this a game while driving in the car, over dinner, or at bedtime. While watching television, the practice could happen during commercial breaks, and students can even challenge their parents to see who can state the best strategy for the fact before solving. This will be a subtle reminder to parents that we are focused on student thinking and that speed will come later. Doing this will also remind parents of the strategies so that they may prompt students when they get stuck. For example, if their child was struggling with 6×7, parents could remind them that they can use their 5's and just add one more group of 7. Isn't that so much better than memorizing, skip counting, or guessing? Hopefully, these little interactions will help parents to see that our goal is not for students to just know discrete facts but that we are trying to build their muscle so that they can generalize across all types of problems and feel confident in doing so.

Parents may need a gentle nudge to reflect upon how committed they are to the homework process. Do they really put in the suggested amount of time each night? Do they show their child that they believe that practice is important by taking the time to play games or ask questions? If they are unable to do this themselves due to work or other family commitments, do they find ways to ensure that it is happening and that their child knows they view it as important? In addition to those questions, it is helpful for you to solicit feedback from parents. Ask them questions about where they think their child is struggling and welcome any suggestions that they may offer. This is so easy for you to do, and you often get really interesting information back. Remember that parents tend to see a whole different child at home than the one you see in your class, and you may find that while Johnny seems like he has it all together, math really stresses him out and he falls apart every night doing homework. You don't have to take parent suggestions per se, but they may have some helpful tidbits that you can use. In our opinion, we don't do this enough, and as a result we miss out on valuable insight into our students that we otherwise may not know.

Depending on the grade level, some parents with concerns may just need to be reassured that their child is fine and that you, the teacher, are not concerned. For example, many parents have unrealistic expectations of what their kids should know and do, and sometimes just knowing that you are comfortable with their progress is enough to ease their worries. This certainly applies to many first-time parents, and you've probably seen it when an older sibling may have made progress more easily, leaving parents concerned that their younger one is somehow behind. Investing in a brief phone call or face-to-face conversation is highly effective at calming parents' fears, but in a pinch, an email works too. That said, if you follow the aforementioned advice about maintaining ongoing communication, this should not be an issue for you.

If you are the one with concerns, ask yourself what they are based on. If strictly going by a timed test, we would caution you about making that the only basis of your evaluation. It is critical to interview students, collect data from daily interactions in class and observations while working in small groups and with partner (Newton, 2014; Sammons, 2019). Get to know how they are thinking and how they respond without the pressure of time. If you have concerns, it is ideal to administer a Math Running Record, since it will give you terrific insight into what strategies students are using, what facts they know, what facts are tricky, and how they are feeling about math. Having that collection of information allows you to have a more complete picture of where the student is and what intervention may be needed. If after you have a more thorough picture you still have concerns, involve parents in the discussion and take stock of resources you may have available in your school such as teacher assistants, math interventionists, or parent volunteers who can help out in the classroom. Having the extra set of hands may help to create more one-on-one opportunities for intervention. If those resources are not an option, consider employing a workshop model so that you will have more time to meet with students in small guided math groups and create space for personalized, purposeful practice in the form of workstations.

HOMEWORK

Let's have a serious conversation about homework. It should be engaging, student-centered, academically rigorous practice. Fluency homework that focuses on multiplication should include the what and the why. We should explain what the trajectory is for learning multiplication and why we teach it this way. We should send home the HOW. Send home flashcards and games that are in the student's zone of proximal development. Which means if John has yet to learn 5 × 8, why on earth would he be working on 5 × 18? Here is a sample letter and some types of activities to send home:

Dear Families,

We are studying our basic math facts. Research shows that learning math facts includes 4 elements; **accuracy, flexibility, efficiency and automaticity.**

Accuracy: We want all of our students to know the correct answer. They learn the facts by practicing them through a variety of activities.
Flexibility: We will be playing several games to build flexibility. We want students to have number sense. We want them to think about and be flexible with numbers. This means that we will work on our facts through strategies rather than rote memorization.
Efficiency: Efficiency means that students can pick quick ways to arrive at the answers based on the numbers that they are using.
Appropriate Strategy Selection: Being able to look at expressions and think about what is the appropriate strategy given those numbers.

Your child is working on Multiplying by 2 **Flashcards**. In this pack you will find 2 things to practice:
1) **Flashcards**
2) **Game**

FIGURE 8.13 *Sample Letter*

PARENTS AS PARTNERS

It's really important to pause and ask yourself how you see the parents of your students. Hopefully, you see them as valuable partners in the education of the students you have been entrusted with. We all have a lot on our plates, and since there never seems to be enough hours in the day, parents can really be a secret weapon in helping to achieve your goals. Of course, this demands that we educate them accordingly and show them that they are a valuable asset to you and a critical partner on the learning journey. If you demonstrate that you appreciate their limited time by providing feedback and support that allows homework time to feel productive, they will feel that their contributions are valued. There is nothing worse than having to struggle for hours at night with your child after a long day and feeling helpless and stressed when they are stuck. When tears can be replaced with laughter, you have succeeded in making parents a true member of the team, and they will be grateful for your efforts to empower them. While not all parents are able or willing to be involved as much as you may like, it's in your best interest to try to get as many parents in your camp as possible and to have alternatives for those who cannot give as much as we would ideally hope.

In making this happen, do not underestimate the power of technology. A quick email here and there really can make all the difference in getting parents off the bench and into the game. A link to a web site with a few photos or videos related to strategies and what fluency looks like, a list of apps or games, suggested web sites, and even suggestions for bedtime stories are little things that mean a lot. Having students create videos to "teach" their families is a double win as students get the additional practice and feel like the experts with their families.

If this seems like a big job, enlist your peers, seek the support of coaches or specialists, and ask your students what ideas they have. Don't be afraid to share the work. It doesn't matter *how* you do it, just *that* you do it! Remember that if you don't take a little time to teach your parents, you could spend *lots* of time undoing the well intentioned "teaching" that they do to help at home. I think that many of us have learned this the hard way!

KEY POINTS

- Home–school connection
- Sharing strategies
- Ongoing feedback
- Addressing concerns
- Parents as partners

SUMMARY

As teachers, we often say that one of our biggest challenges in getting students to mastery is time. There are so many standards and so little time, and students often come to us lacking mastery of critical prerequisite skills. This makes our job bigger, and we can definitely use all of the help we can get!

While parents likely won't be able to support all of the math we teach, they can play an important role in the acquisition of fluency with basic math facts. Whether that role is productive or destructive depends on us. We need to ensure that parents can support their children in ways that foster flexibility, honor thinking, and deemphasize speed as an indicator of success. The home–school connection must be cultivated in intentional ways so that parents are empowered to reinforce the experiences that students are having at school. We hope that you will use the ideas presented here to open the lines of communication with parents and truly engage them as partners in this process. Keep in mind that many may not have fond memories of math, and some may even be math-phobic. This may mean it will take some convincing, but if you stick with it and help them to understand why and how you approach this, you will get a great return on your investment!

REFLECTION QUESTIONS

1. What are you currently doing around fact fluency, homework, and parent support?
2. What does your current communication with parents around fluency look like?
3. What new ideas from this chapter do you have?

CALL TO ACTION

 1. Share your favorite "Aha!" moment about parental involvement with fact fluency on social media to help spread the movement! #FDJH

 2. Take a photo of different ways that you are working to involve parents with basic fact fluency and share it on social media to encourage other teachers to do it too! #FDJH

3. Get started with this "make your own game board" as a homework activity! Families can make their own basic fact fluency board games to play at home, focusing on what their child needs to practice!

REFERENCES

Baroody, A. J. (2006, August). Why children have difficulties mastering the basic number combinations and how to help them. *Teaching Children Mathematics, 13*, 22–31.

Entwisle, D. R., & Alexander, K. L. (1996). Family type and children's growth in reading and math over the primary grades. *Journal of Marriage and Family, 58*, 341–355.

Epstein, J. L. (2001). *School, family, and community partnerships: Preparing educators and improving schools.* Westview Press.

Gill, S., & Reynolds, A. J. (1999). Educational expectations and school achievement of urban African American children. *Journal of School Psychology, 37*, 403–424.

Halle, T. G., Kurtz-Costes, B., & Mahoney, J. L. (1997). Family influences on school achievement in low-income, African-American children. *Journal of Educational Psychology, 89*, 527–537.

Holloway, S. (1986). The relationship of mothers' beliefs to children's mathematics achievement: Some effects of sex differences. *Merrill-Palmer Quarterly, 32*, 231–250.

Hurst, C. (2015). The multiplicative situation. *Australian Primary Mathematics Classroom, 20*(3), 10–16.

Malola, M., Symons, D., & Stephens, M. (2020). Supporting students' transition from additive to multiplicative thinking: A complex pedagogical challenge. *Australian Primary Mathematics Classroom, 25*(2), 31–36.

Newton, N. (2014). *Guided math in action: Building each student's mathematical proficiency with small-group instruction.* Routledge.

Sammons, L. (2019). *Guided math: A framework for mathematics instruction second edition.* Teacher Created Materials.

Shaver, A. V., & Walls, R. T. (1998). Effect of Title 1 parent involvement on student reading and mathematics achievement. *Journal of Research and Development in Education, 31*, 90–97.

Sheldon, S. B., Epstein, J. L., & Galindo, C. L. (2010). Not just numbers: Creating a partnership climate to improve math proficiency in schools. *Leadership and Policy in Schools, 9*(1), 27–48.

Starkey, P., & Klein, A. (2000). Fostering parental support for children's mathematical development: An intervention with Head Start families. *Early Education and Development, 11*, 659–680.

Van Voorhis, F. L. (2007). Can math be more meaningful? Longitudinal effects of family involvement on student homework. Paper presented at *the annual meeting of the American Educational Researchers Association.* Chicago, IL.

CHAPTER 9

Action Plan

Every school should have a mission statement and vision of fluency that everyone who is a stakeholder understands and can articulate. A mission statement is a short written statement about your goals and philosophies about math fluency in your school. The mission statement should discuss what your school does to develop mathematicians, how it does it, and why it does it. It is really an expression of your values around teaching and learning math. In order to get a clear understanding of the mission around fluency, you should have teachers submit a quick statement either on a piece of paper or through an online survey system like Survey Monkey or Google Forms. Then, discuss the statements and come up with a shared mission statement. It should be clear, concise, specific, and useful so that everyone can understand and execute it (see Figures 9.1–9.3).

After you know what the mission is, then you can create a vision statement that describes exactly what it looks like. Your vision statement should be short, definitely under 20 words. It should clearly communicate what you are striving toward as a school in a way that everyone (all stakeholders, secretaries, custodians, parents, students, teachers) understands, can remember, and can then tell others about.

After you have a vision statement, then you can create a fluency plan. A fluency plan articulates the exact steps that you are going to take to achieve the mission and reach the vision. Schools need a fluency plan that guides teachers both horizontally and vertically. In this chapter, we are going to look at some templates to help you plan. The mission statement and vision must be communicated to and shared with all stakeholders, and each one should clearly understand their role. Here is an example of a mission statement, a vision statement, and a fluency plan:

Mission Statement:

Our school's math goals are to foster the growth of mathematically minded students. We want to provide an academically rigorous, standards-based, engaging, real-life-connected curriculum that inspires and encourages students to do their very best and enjoy mathematics. We aim to provide high-quality instruction that allows all students to experience the joy of success and the love of being a life-long learner who sees math in action in their everyday lives.

DOI: 10.4324/9781032614229-9

Fluency Vision Statements:

- To make math enjoyable, engaging, and achievable for all students!
- That all students at our school will be fluent with their basic math facts!
- That all students at our school will achieve grade-level math fact fluency!
- Every student in our school is given the scaffolding they need to achieve basic fact fluency!

GETTING ALL THE STAKEHOLDERS ON BOARD

1. Does your school have a fluency vision? Is there a fluency mission statement?	2. Is everyone at your school operating from the same understanding of fluency?
3. What are you doing so that parents can effectively help their students develop fluency including accuracy, flexibility, efficiency and automaticity.	4. What are you doing to empower all your students around ideas of being and becoming fluent?

FIGURE 9.1 Questions to Get You Started

Mission Statement: The mission statement should discuss what your school does to develop mathematicians. 1) How it does it: 2) Why it does it: 3)What are the underlying values:	**Vision Statement:** It describes exactly what it looks like:
Fluency Plan: What are the first steps: Step 1) Step 2) Step 3)	What are the easy parts of creating these? What are the challenges of implementing these ideas?

FIGURE 9.2 Big Picture

FLUENCY PLAN

Understanding Basic Facts				
Grade	**Beginning of the Year September**	**1st Marking period November**	**2nd Marking period February**	**End of the Year**
3rd Grade	Checking for Understanding of Equal Groups	Students should understand multiplying by 0,1,10 and 5	Students should understand multiplying by 2,4,8 and then 3,6,9	Students should know and understand all facts including 7's
4th & 5th	Give the students a Math Running Record	Review 3,6 and then teach 12's Review 10s and teach 11s	Use helper facts to explore 1 digit by 2 digit	Continue to use helper facts and properties to work on 2 digit and 3 digit multiplication and division by 1 digit

FIGURE 9.3   Example of a Fluency Plan

1. Where are your students right now?	2. What evidence do you have of where they are?
3. What do you need to do to know where they are?	4. Do your assessments address all 4 elements of fluency: accuracy, flexibility, efficiency and automaticity.

FIGURE 9.4   Fluency Plan Template

QUESTION 1: WHAT IS YOUR FLUENCY PLAN?

1. What are you doing right now to assess on a regular basis?	2. How do you keep track of your ongoing assessments?
3. What types of exit slips do you do that assess fluency?	4. What sort of system do you have so that students can track their own individual progress?

FIGURE 9.5 Fluency Plan

QUESTION 2: WHAT ARE YOU DOING ABOUT ONGOING ASSESSMENT?

There is a much deeper dive that you can take with your school on www.mathfactfluen-cyplayground. Administrators and teachers can use this template to take a deep dive into thinking about the five components: Research Framework; Getting the Data; Classroom Design; Teaching for Fluency; and then Reflecting and Revising the Plan. Here is an overview of that template. There are many other different sections on the site to guide the fluency discussion.

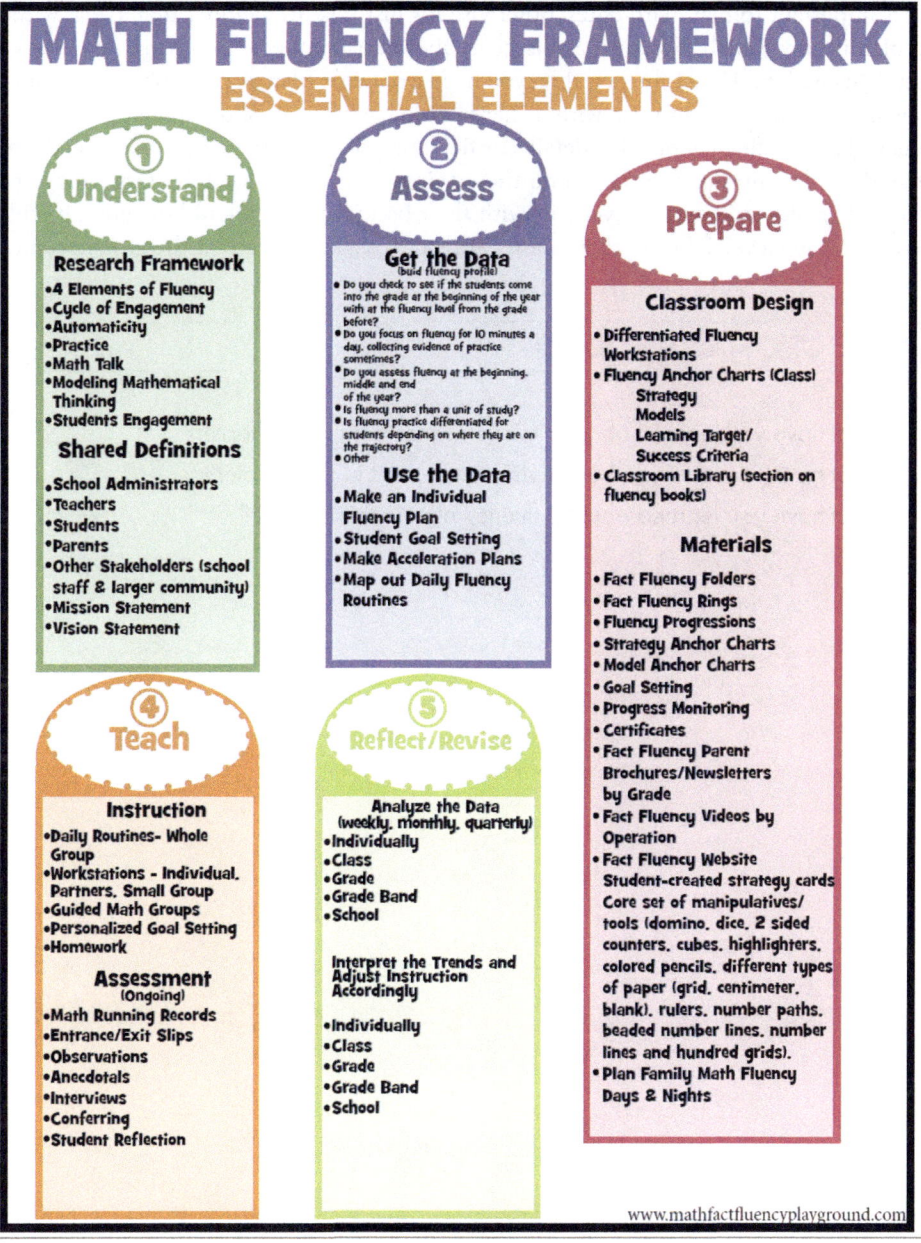

KEY POINTS

- Mission statement
- Vision statement
- Fluency plan

SUMMARY

It is important to plan with dates; otherwise it's just talk. Remember that a goal without a plan is only a wish! So, if you are serious about raising the achievement levels of your students with basic fact fluency, then you need a detailed plan. A plan starts with a shared mission statement, continues with a shared vision statement, and culminates with a fluency plan. The fluency plan is a detailed articulation of what is going to happen and when, based on the current data. If you do these things, you will absolutely start a movement toward greater student achievement with their basic facts! Think of how great it will be if every grade level achieves their mastery objectives and you can hit the ground running!

REFLECTION QUESTIONS

1. What have you learned about a mission statement in this chapter?
2. What have you learned about a vision statement in the chapter?
3. What have you learned about a fluency plan in this chapter?

CALL TO ACTION

 1. Share your favorite idea from the entire book on social media! #FDJH

 2. Take a photo of something you want to share to encourage others to use games, stories, energizers, and routines to explore basic fact fluency! #FDJH

3. Be sure to stay in touch!